A Celebration of 50 Years of Progress in Biological Standardization and Control at WHO

Developments in Biological Standardization

Vol. 100

This series Dev Biol Stand begins with Vol. 23 and is the continuation of both «Progress in Immunobiological Standardization, Vols 1-5» and «symposia series in Immunobiological Standardization, Vols 1-22».

KARGER

Basel · Freiburg · Paris · London · New York ·
New Delhi · Bangkok · Singapore · Tokyo · Sydney

A Celebration of 50 Years of Progress in Biological Standardization and Control at WHO

WHO Headquarters
Geneva, 25-27 October 1998

Volume Editors

Fred Brown
Plum Island Disease Center, Greenport NY, USA

Elwyn Griffiths
World Health Organization, Geneva, Switzerland

Florian Horaud
Institut Pasteur, Paris, France

Geoffrey Schild
National Institute for Biological Standards and Control, Potters Bar, U.K.

Proceedings of a Symposium organized and sponsored by
the International Association for Biologicals (IABS) and
the World Health Organization (WHO)

24 figures and 26 tables, 1999

KARGER

Basel · Freiburg · Paris · London · New York ·
New Delhi · Bangkok · Singapore · Tokyo · Sydney

Developments in Biological Standardization

Bibliographic Indices. This publication is listed in bibliographic services, including Current Contents® and Index Medicus.

Drug dosage. The authors and the publisher have exerted every effort to ensure that drug selection and dosage set forth in this text are in accord with current recommendations and practice at the time of publication. However, in view of ongoing research, changes in government regulations, and the constant flow of information relating to drug therapy and drug reactions, the reader is urged to check the package insert for each drug for any change in indications and dosage and for added warnings and precautions. This is particularly important when the recommended agent is a new and/or infrequently employed drug.

Printed in Switzerland by Médecine et Hygiène, Genève
ISBN 3-8055-6952-1

Meeting Coordinators

E. Griffiths (Geneva, Switzerland)
F. Horaud (Paris, France)
J. Petricciani (Palm Springs, CA, USA)
G. Schild (Potters Bar, U.K.)

Organization of the Symposium

Opening Session:	Historical Perspectives	
Session I:	Achievements of the First 50 Years of Biological Standardization at WHO	
	Co-chairs:	*Johannes Löwer (Germany) and*
		Elwyn Griffiths (Switzerland)
Session II:	Prospects for Standardization in the 21st Century	
	Co-chairs:	*Alan Shaw (USA) and*
		Takao Hayakawa (Japan)
Session III:	Future Challenges	
	Co-chairs:	*Kathryn Zoon (USA) and*
		Victor Grachev (Russia)
Session IV:	International Perspectives	
	Co-chairs:	*Geoffrey Schild (U.K.) and*
		Ian Gust (Australia)

The Meeting Coordinators of

A Celebration of 50 Years of Progress in Biological Standardization and Control at WHO

are grateful to the following companies and organisations for their financial support

- Ares Serono (Switzerland)
- Chiron Vaccines (USA)
- CSL (Australia)
- Interpharma (Switzerland)
- Medeva Pharma Limited (U.K.)
- Merck & Co. Inc. (USA)
- Pasteur Mérieux Connaught (France)
- Pasteur Mérieux MSD (France)
- SmithKline Beecham Biologicals (Belgium)
- Swiss Serum & Vaccines Institute (Switzerland)
- Wyeth-Lederle Vaccines & Pediatrics (USA)

Contents

Brown F, Griffiths E, Horaud F, Schild GC (eds): A Celebration of 50 Years of Progress in Biological Standardization and Control at WHO.
Dev Biol Stand. Basel, Karger, 1999, vol 100, p 1

Opening Session

Brown F, Griffiths E, Horaud F, Schild GC (eds): A Celebration of 50 Years of Progress in Biological Standardization and Control at WHO.
Dev Biol Stand. Basel, Karger, 1999, vol 100, pp 3-4

Welcome Address

F. Horaud

President, International Association for Biologicals (IABS)

Madam Director-General, Doctor Mérieux, Ladies and Gentlemen, Colleagues and Friends,

It is a privilege and pleasure for me as President of the IABS to welcome Dr. Brundtland, Director-General of WHO, and all of you to this historic meeting to celebrate 50 years of progress in biological standardization which goes back, as we will hear, to Paul Ehrlich and Sir Henry Dale at the beginning of this century. The importance of Biological Standardization cannot be overstated. It is recognised as a normative activity in the constitution of WHO. The foundations in this field laid by the League of Nations and continued by the WHO underpin many of the major activities of WHO. Most developments in prophylaxis and the therapy of infectious diseases are based on standardized biological products whose potencies are traced directly to WHO International Biological Standards. The quality and safety of vaccines, blood products and other biologicals are dependent on the requirements for these materials prepared by WHO on the advice of experts from many countries.

The eradication of smallpox and the expected eradication of polio would have been impossible without the availability of vaccines of defined, consistent and guaranteed quality, produced by the pharmaceutical industry and national authorities to requirements originating from WHO.

During the past fifty years, many changes have taken place in the range, nature and use of biological products. This period has seen the development of antibiotic therapy as a routine procedure and the introduction of polio, measles, hepatitis B and other vaccines, in the process saving millions of lives and raising the quality of life of millions more. We have seen the new era of recombinant products take an established place, offering continued and greater availability of products for replacement therapy, such as insulin, erythropoietin, growth hormone, coagulation factors and an increasing number of biologicals not previously available from natural sources in sufficient amounts. We have a better understanding of viral contamination of blood and how to minimise its risks. There have been many challenges and many achievements.

The International Association for Biologicals since its foundation in 1955, has been privileged to collaborate closely with the World Health Organization. Our Association always considered this action a priority since biological standardization is the final result of a consensus emerging from an international scientific debate.

Many scientific symposia and congresses have been organized jointly by WHO and IABS, and several of them are milestones in biological standardization. IABS and WHO have the same objectives in securing universally applied high standards of quality, safety and efficacy of biological products used in world medicine; a single high standard for the world. WHO and IABS have formed a strategic alliance with the objective of consensus. Links between the IABS and the Biologicals Unit of WHO, currently under the leadership of Dr. Elwyn Griffiths have been highly constructive.

We are here both to celebrate the achievements of WHO in biological standardization over the past 50 years and to look to the future and see what can be done. IABS offers its full support to WHO in the continuing efforts to provide an authoritative, global lead in biological standardization. IABS congratulates WHO on 50 years of achievement in the field of biological standardization and looks forward to sharing with WHO the next 50 years of successful progress in this field which is a crucially important aspect of world medicine.

Prof. F. Horaud, Institut Pasteur, 28, rue du Docteur Roux, F-75724 Paris Cedex 15.

Brown F, Griffiths E, Horaud F, Schild GC (eds): A Celebration of 50 Years of Progress in Biological Standardization and Control at WHO.
Dev Biol Stand. Basel, Karger, 1999, vol 100, pp 5-7

Opening Address

G.H. Brundtland

Director-General, World Health Organization, Geneva, Switzerland

Let me first thank Professor Horaud, President of the International Association for Biological Standardization for inviting me to this reception, and likewise let me welcome members of the International Association for Biological Standardization and their guests to the Headquarters of the World Health Organization.

As Professor Horaud has just mentioned, WHO has been setting standards for biological medicines for over half a century and is justly proud of its achievements in this field. We all know that biological products, which include vaccines, blood products and biological therapeutics, are central to the achievement of the WHO's mission «Health for All». However, the very nature of biologicals raises particular and common questions regarding their quality control, and the considerable hazards associated with some of these substances need continuous vigilance.

Internationally, the importance of biological standardization for global health programmes was recognized in the early years of this century and the provision of international biological standards was developed by the League of Nations through its Permanent Commission on Biological Standardization. However, this work was interrupted by the Second World War and one of the first actions of the Interim Commission of the World Health Organization in 1946 was to re-establish international activities in this area by setting up the Expert Committee on Biological Standardization.

Although 1998 marks the 50th Anniversary of the World Health Organization, the Expert Committee on Biological Standardization celebrated its 50th Anniversary last year, since its first meeting was in June 1947. Technically, it precedes WHO itself, indicating the importance given to biological standardization by the embryonic WHO. There is hardly any need for me to remind you of the success of WHO's biological standardization since that time, nor of its importance for both developed and developing countries. Through its Biologicals Unit, which coordinates the work of the Expert Committee on Biological Standardization, and that of the network of WHO International Laboratories for Biological Standards, WHO has played a key role in establishing the international reference materials needed for calibrating biological medicines and in developing recommendations on their production and quality control to ensure safety and efficacy.

The success of the smallpox eradication programme and the more recent advances towards eliminating polio worldwide have been possible because of the availability of safe and effective vaccines. It is not an exaggeration to suggest that in the absence of an international framework for the standardization and control of these vaccines and other biologicals which have been controlled using WHO standards and technical guidance, progress in the control of infectious diseases would have suffered enormously. Likewise WHO, through its biologicals programme, has played a crucial role in standardizing antibiotics worldwide, which has resulted on the successful treatment of hundreds of millions of infections. In addition, WHO provided both physical standards and written guidance for the manufacture and control of safe and effective blood products that are required in numerous life-threatening situations.

Biologicals – Increasing Complexity and Diversity

In recent years, the field of biologicals has been one of expansion and increasing diversity. In some instances traditional products are being replaced by material derived by recombinant DNA technology. This can lead to potential standardization problems as encountered with, for example, interferons and plasma products. Also, new possibilities for diagnostics are emerging such as the use of gene amplification methods for the virological safety testing of blood and blood products and there are exciting new approaches to control testing with the possibility of molecular based techniques leading to a reduction of testing in animals and perhaps to an increased safety margin for the products.

Even so, new technologies can lead to new issues, such as the recent discovery using a highly sophisticated and sensitive molecular technique, of low levels of reverse transcriptase activity in chicken cell-derived vaccines. Here difficult decisions have had to be made about the significance of this discovery for vaccine safety, using the best scientific data available.

The increasing complexity and sophistication of biologicals, as well as their growth in volume, presents a considerable challenge, especially for the developing world. Developing countries can be expected to benefit greatly from the use of new biologicals, such as vaccines, and some are developing their own capacity in the new and highly sophisticated biotechnologies. The need for a coordinated international approach to the standardization and control of these substances is therefore, increasing and its importance cannot be over emphasized.

Of course, the ultimate responsibility for ensuring the safety and efficacy of biological medicines rests with governments. However, some national control authorities are inexperienced in dealing with biologicals whether derived by traditional or novel biotechnologies. Others lack expertise and resources. The strengthening through education, training and updating, where appropriate, drawing upon the knowledge and skills of those authorities already experienced in this area, will be necessary in the future, and regional collaboration needs to be promoted and strengthened.

World Health Assembly – Resolution

Recognizing these developments and needs, the World Health Assembly in May 1997 unanimously adopted a resolution on the Quality of Biological Products Moving in International Commerce. The resolution recognizes the special technical expertise needed for evaluating and controlling biological products, as well as the long-standing and valuable role of WHO's Biologicals Unit and the Expert Committee on Biological Standardization. However, it also recognizes that WHO's standardization activities need to be strengthened to meet the challenges of the rapid expansion and increasing complexities of the biologicals field as it moves into the 21st century.

Following the recommendations of the Resolution, an independent review of WHO's activities in biological standardization was commissioned with a view to recommending action that will enable WHO to respond to scientific developments in a timely manner and to strengthen the mechanism for providing clear norms and active leadership in promoting the safety, quality and efficacy of biological and biotechological products used in medicine. This review is now completed and the report of the external review team, which consulted widely, is expected soon.

Restructuring at WHO

Considerable changes in the organization of WHO are currently underway. At Headquarters we have grouped the programmes into nine clusters, each sending a clear message of what business we are in. In the coming months, under the supervision of the Executive Directors, each cluster will be streamlining its activities to optimize what we can do together across the whole organization and in particular with other partners.

What we do in Geneva matters very little if it does not have an impact in Member States. Clearly, WHO's Biologicals Standardization Programme has made a big impact over the years and we must ensure that it continues to make a difference for health in the coming years. The recommendations of the independent review will need to be considered during this restructuring phase.

Partners

I mentioned the need to have partnership with others. An ideal partnership already exists in the biologicals field and WHO values very much the long standing support given to it by the International Association of Biological Standardization. The IABS, which is a Commission of the International Union of Microbiological Societies, a non-governmental organization in official relations with WHO, brings together manufacturers, regulatory personnel, research workers from academic institutions and scientists from public health organizations and has done much to further improvement in the safety and quality of biological substances used in medicine. I know that you promote biological standardization by organizing meetings on subjects of contemporary importance in the biologicals field and by pu-blishing the proceedings of the international meetings on developments in biologicals standardization and in the major international journal, Biologicals. In this way, you provide a much-needed platform for discussing the many issues related to the production and control of safe and effective biologicals, and on which WHO can draw for the best scientific views and concerns.

On this occasion you have organized a symposium to celebrate the success of WHO in this field – 50 years of progress in biological standardization and control at WHO – and for this we are indeed extremely honoured. I hope that WHO and the IABS will go on working together in this way into the next century to ensure that the biological medicines we need for our global health programmes are safe, are good quality and are effective.

Mrs. J. Dyrhauge, Office of the Director-General, World Health Organization, 20, avenue Appia, CH-1211 Geneva 27, Switzerland.

Brown F, Griffiths E, Horaud F, Schild GC (eds): A Celebration of 50 Years of Progress in Biological Standardization and Control at WHO.
Dev Biol Stand. Basel, Karger, 1999, vol 100, pp 9-10

The IABS Supporting the Work of the World Health Organization The Eradication of Poliomyelitis

C. Mérieux

Fondation Marcel Mérieux, Lyon, France

The International Association of Biological Standardization was founded in 1955 in Lyon, when we took the bold initiative of organizing a European symposium to bring together public and private vaccine manufacturers, state regulators, researchers and, of course, veterinarians. The World Health Organization was represented by Niels Jerne.

When we decided to include polio immunization on the agenda, we had no idea that the tragic accidents in California would make our meeting in Lyon so very timely. Our colleagues from the United States were with us in Lyon. They found themselves in the presence of an exceptional audience. The contacts they established with a group of Russian virologists were perhaps the source of their co-operation on the Sabin vaccine, while in the United States the enormous programme against poliomyelitis used only the Salk injectable vaccine. In any event, it was very clear that from the outset our Association would base its sense of unity on, and prove its effectiveness with, the standardization of the two poliomyelitis vaccines.

The injectable, inactivated polio vaccine developed by Jonas Salk was initially produced in Canada. A desire to speed up production is probably what led to a breakdown in the controls at a large American laboratory, and the subsequent tragedy in San Francisco. In 1955, the Salk Vaccine was not produced in Europe. Public Institutes in the Netherlands and Scandinavian countries intended to produce it and eventually manufactured excellent vaccines to eradicate poliomyelitis in their respective countries.

We began producing Pierre Lepine's French vaccine in conjunction with the Pasteur Institute and, in the 1970s, as part of our vaccinology collaboration, we followed the example of Jonas Salk and stopped using monkey cells to produce the vaccine, replacing them by Vero cells. We also worked very openly with the National Institute of Health, especially with John Petricciani, and we collaborated in the same way with the Dutch Institute at Bilthoven. We used the Dutch technique developed by Van Wezel to achieve truly industrial production; every one thousand litre tank produced many thousands of vials. This explains why, in France, we have continued to use the injectable polio vaccine in combination with DTP. One can only hope that now, on the eve of total eradication of poliomyelitis all over the world, the use of combined vaccines will become more and more prevalent for faster eradication.

In 1960, a new European meeting was organized in Lyon to discuss the oral live polio vaccine. Albert Sabin attended the meeting. Subsequently, Behringwerke, Wellcome, and Sclavo concentrated exclusively on the Sabin vaccine, as did RIT, later to become SKB. Naturally we must not overlook the important role played by our Russian colleagues and should pay homage to our friend Chumakov.

Our laboratories manufactured the Sabin vaccine for export. You are probably all aware that, in 1979, Alain Mérieux gave the type 3 strain, developed by Robert Stones, to the other manufacturers free of charge. This attests to the spirit the IABS managed to create, not only between vaccine manufacturers and regulators, but above all with the World Health Organization and Frank Perkins. Thanks to this spirit of mutual trust, the Sabin vaccine has played a key role in the World Health Organization's upcoming victory over polio, following the triumph of the smallpox vaccine. I should point out that today these two live vaccines would probably not be licensed, especially in this age of AIDS anxiety, followed by the panic about British cows, which triggered a spell of delirium regarding safety issues and formidable pressure from the media. It is becoming increasingly difficult to develop new drugs, and perhaps impossible to create new vaccines.

We were nevertheless filled with fresh enthusiasm in 1995, at the conclusion of the year commemorating Pasteur, at the Vaccinology Congress held in Marnes-la-Coquette, where Pasteur once worked. The President of the Sabin Foundation, in the company of Mrs Sabin and Darell Salk, came especially to pay solemn tribute to Jonas Salk. They recalled the epic of polio immunization and expressed their wish that this worldwide effort would serve as a model in the fight against AIDS. The Sabin Foundation joined with the Rockfeller Foundation in saying: «Against AIDS, we have neither means nor the right to continue to apply the industry's traditional answers».

I think that it is up to IABS to envisage this kind of action for the future. It is extraordinary to observe that, forty years after the Association was created, it still has the same style and the same pace. Today's meeting is proof that it is still vested with the trust of the World Health Organization.

Dr. C. Mérieux, Fondation Marcel Mérieux, 17, rue Bourgelat, F-69002 Lyon, France.

Brown F, Griffiths E, Horaud F, Schild GC (eds): A Celebration of 50 Years of Progress in Biological Standardization and Control at WHO.
Dev Biol Stand. Basel, Karger, 1999, vol 100, pp 11-15

H.H. Dale and the London Centre for WHO Standards Some Milestones of the Early Years

D.R. Bangham

Mill Hill, London, U.K.

Key words: HH Dale, history, biological standardization.

Abstract: Henry Dale, a Nobel laureate and statesman of science, helped to organize the meetings in 1923 and 1925 to set up international standards for insulin and other biologicals. He made the National Institute for Medical Research one of the two world centres for standards. Some milestones in the work of the London centre are described: (i) the first standards; (ii) vitamins and hormones; (iii) WHO standards for many antibiotics; (iv) the provision of an international working standard for ACTH; (v) the old and new methods of ampouling; (vi) the impact of research and immunoassay on the need for standards; and (vii) the special ECBS sessions for endocrinology and haematology. Dale tells how he had to preserve the first batch of insulin for use as the standard instead of for treatment.

On this occasion, at which the pioneers of international biological standards are honoured, **Henry H. Dale** of London stands as a pre-eminent pioneer. Dale was a great statesman of science: Nobel laureate in physiology, President of the Royal Society and Director of the National Institute for Medical Research (NIMR). The key to his seminal and sustained support for biological standards lies, I believe, in his early responsibilities for 10 years in the Wellcome Physiological Research Laboratory. As Director he was responsible for the potency and safety of diphtheria and tetanus antitoxins and organic arsenical products made by Burroughs Wellcome. He gained further stimulus for it by a visit to Ehrlich's laboratory. He joined the Medical Research Committee in 1914 and in 1920 helped to draft the first principles for legislated control of drugs in Britain.

In 1921 Dale visited Banting and Best and foresaw the huge potential benefit of **insulin**. At T. Madsen's invitation he wrote a strong paper stating the need for international standardization of biologicals, which instigated the first meeting to discuss the subject at the Physiology Congress in 1923. Significantly, he offered the NIMR as a research centre for «non immunological» biological standards and assigned P. Hartley to look after it. T. Madsen's Statenserum Instituut (SSI) in Copenhagen was appointed to look after microbial «immunological» standards.

In 1925 Dale chaired the first session of the Permanent Commission on standardization of the **League of Nations** (LON). First International Standards for insulin (which had been prepared at NIMR), posterior pituitary, digitalis, Salvarsan and diphtheria and tetanus antitoxins were established. At subsequent annual sessions, standards for various microbial antisera prepared in Copenhagen or London were established. During the 1930s standards for several **vitamins** and for **steroid hormones** were set up by the NIMR, many of which came from current research in Britain.

Dale also chaired the meeting of the Interim Commission at the end of World-War II, in 1944 in London, at which the IS for **Penicillin** was established. He insisted that its potency should be defined in international units (IU), not «microgram equivalents» then used in the USA. AA Miles, who succeeded Hartley, by a strategical paper given at the 3rd ECBS, ensured that the standard for streptomycin and other later antibiotics also had IUs.

During the 1950s, Niels Jerne, a Dane, later a Nobel laureate in Physiology or Medicine, exerted a strategic influence on WHO standardization. In the early 1950s he represented the SSI at WHO committees, and from 1956-60 he was Chief of Biologicals at WHO. He developed certain principles of bioassays; he instigated the production of **WHO Requirements** for manufacture, those for smallpox, yellow fever and polio vaccines, and for sterility, were the earliest; and he stimulated WHO to form the Division of Immunology in the 1960s.

At the NIMR, Miles was succeeded by WLM Perry (1952-58) and DG Evans (1958-60) in charge of the Standards Division. I succeeded Evans until 1972 when the MRC Division was absorbed into the new National Institute for Biological Standards and Control (NIBSC). In 1960, AS Outschoorn (1960-75) succeeded Jerne as Chief at WHO.

In 1957, **ACTH**, a newly discovered hormone, was then of intense interest, but it was impossible to standardize batches for clinical use. This problem could only be tackled if every batch was assayed against the same standard. This required the provision of huge numbers of ampoules of an international working standard (IWS), each ampoule containing identical but very small amounts of the same material as the IS. Until 1958 every glass ampoule of a standard was made, filled and sealed by hand – clearly unworkable for the many ampoules needed for an IWS and the increasing numbers of standards. In 1960, a whole **new ampouling procedure**, including freeze-drying, was developed at NIMR. Special equipment was designed and built there, and continued in use for 25 years. Output of ampoules for standards reached 100,000 per year; 38,000 ampoules of the IWS for ACTH were used worldwide in the next 20 years. Until freeze-drying was used for standards, the IU was defined as a certain weight of the standard. With freeze-drying applied, a stated number of IUs was assigned to the ampoule content for each standard.

In the 1960s and 1970s, WHO reference materials for some 80 **antibiotics** were established by WHO. Most of these were set up by JW Lightbown, at NIMR until 1972, and then at NIBSC. WW Wright (USA) was a notable helper.

By the 1960s **research in human physiology** and pathology was developing explosively, and there was clear need for standards for substances involved in such research. To some extent this need was met by 'MRC Research Standards' from the London laboratory and prepared with expert research work worldwide. Although of no official status, they were used worldwide. The introduction in the early 1960s of **radioimmunoassays** gave a tremendous stimulus to research. The ECBS in 1966 discussed the mechanism, meaning and validity of this novel assay methodology. The Committee in 1968 decided that standards, assays and units by immunoassay systems should be distinguished from those for bioassay estimates. The impact of the new methodology was widespread.

In 1972, the work and staff of the NIMR Division was transferred to the **new NIBSC**, «temporarily» housed in the old NIMR building at Hampstead. DG Evans was Director until 1976 when, after more than 50 years, the MRC handed responsibility for standards to the Board of NIBSC. JWG Smith was Director until 1985 when GC Schild was appointed. In 1987 NIBSC moved to its new extensive buildings at South Mimms.

In '1973 exhortation at the World Health Assembly by the International Society for Endocrinology, on behalf of assays and WHO standards needed for hormone research, was acknowledged by WHO. The ECBS in 1974 was devoted entirely to **endocrinology**. Standards for 14 hormones were established, some of them MRC Research Standards already in use. The first guidelines on quality of immunoassay kits was published in the ECBS Report.

In 1976, with FT Perkins as WHO Chief of Biologicals, a special meeting on standardization in the field of **haematology** was held. Several relevant new standards already set up by London were established.

Thereafter the scope of the ECBS has grown vastly and variously. At WHO, Perkins (1976-84) was succeeded by J Petricciani (1985-88) and D Magrath (1989-1996). **Recombinant proteins** were first considered in 1983. Interferons, cytokines, growth factors and a great range of reference materials for viruses continue to develop. Standardization of microbial immunologicals by Copenhagen, and of veterinary ones by the Central Veterinary Laboratory, Weybridge, have been taken over by NIBSC. Analytical methodology becomes ever more sophisticated. Early standards of substances now so easily characterized by chemical and physical means, have been discontinued. The early era of biological standardization has changed utterly.

While each generation discovers new principles of standardization and its handmaidens **biometry** and **statistics**, it is right that Dale and the early pioneers should be remembered.

A glimpse of Dale's personality can be seen in a short film made in the 1970s. In this he talks, retrospectively, about a clergyman who came in 1923 to beg from Dale some of the small amount of insulin, then at NIMR, to treat his daughter who had diabetes. Dale reluctantly had to refuse this, explaining that the preparation he had was committed to become a standard against which all other preparations would be assayed to determine their strength. Later the daughter died, but the clergyman wrote a letter to Dale saying «thank God for the Medical Research Council». Although the film was made many years later, it shows the old Dale still much moved by that event.

Reference

Bangham DR: A history of Britain's contribution to biological standardization, 1998, pp 300, in press.

Dr. D.R. Bangham, 4 Crown Close, Mill Hill, London NW7 4HN, U.K.

H.H. Dale.

C.R. Harington

P.B. Medawar

P. Hartley

Milestones in Biological Standardization

1923	Physiology Congress, meeting on standards Dale's offer of NIMR as centre, Madsen offers SSI at Copenhagen
1925	Permanent Commission of LON 1st standards for insulin, digitalis, post.pituitary, diphtheria and tetanus antitoxins, Salvarsan
1930s	Standards for vitamins and for steroids, by NIMR
1944	Standard for Pencillin, with IU assigned
1950-70	Other 80 antibiotics, JW Lightbown
1958	Need for IWS for ACTH
1960	new ampouling procedure & equipment designed and built at NIMR Output 100,000 ampoules/year
1955-60	Jerne's contributions: – principles of bioassay – WHO International Requirements – WHO Division of Immunology
1960	Increasing research in h physiology & pathology Impact of immunoassays
1965	MRC Research Standards, human hormones, blood and antibodies
1972	New NIBSC formed at Hampstead
1974	ECBS and session on endocrinology
1976	After 50 years MRC relinquish standards to NBSB
1977	ECBS and session on haematology
1982	Recombinant methods considered at WHO
1985	Steady increase in reference materials, interferons, viruses, cytokines, growth factors, blood components
1987	NIBSC move to South Mimms

Brown F, Griffiths E, Horaud F, Schild GC (eds): A Celebration of 50 Years of Progress in Biological Standardization and Control at WHO.
Dev Biol Stand. Basel, Karger, 1999, vol 100, p 17

SESSION I

Historical Perspectives

Co-chairs: *Johannes Löwer (Germany)*
and Elwyn Griffiths (WHO, Switzerland)

Brown F, Griffiths E, Horaud F, Schild GC (eds): A Celebration of 50 Years of Progress in Biological Standardization and Control at WHO.
Dev Biol Stand. Basel, Karger, 1999, vol 100, pp 19-30

International Biological Standardization in Historic and Contemporary Perspective

M.R. Hilleman

Merck Institute for Therapeutic Research, Merck Research Laboratories, West Point, PA, USA

Key words: Biologicals regulation, World Health Organization and Biologics.

Abstract: Recounting the origination and development of the concepts and institutions to foster and control new and existing biological products is a fertile subject for historical review. It describes what was a necessary and functional activity that needed to be brought into being.
The earliest examples of biologicals regulation were created by the eighteenth century inventors themselves and became progressively institutionalized by the creation of national control authorities which increased in number and sophistication during the past century. In 1948, the World Health Organization (WHO) was established and created biological standards and regulations as an advisory to member nations. This too has grown in sophistication and is now of central importance in the biologicals enterprise worldwide.
The field of biological standardization and regulation can be viewed in its need and role to foster technological development and to assure that the products of that endeavour are worthy and are retained. In many respects, it is a tale of troubles, tragedies, and triumphs and this is recounted in the text.

Introduction

It is a pleasure for me to participate in this 50th Anniversary Celebration of the World Health Organization [1, 2] and that of its Division for Biologics. And, even more, to celebrate the sense and wisdom of the regulators themselves who have done so much to promote biologicals to the present time. The Biologics Unit at WHO has provided yeoman service to the scientific community and to the peoples of the world for five decades. Its dedication to service, together with its outstanding achievements, give it much reason to be recognized and to be proud of what it has accomplished.

I have never served as a biological regulator, but I did join the ranks of the regulated in 1944. I have served as an advisor to the World Health Organization, particularly in biologicals, and to the U.S. biologicals control authority and have participated in the preparation of standards and regulations for many new products. From these activities I have gained insight into the seminal importance of regulatory control to the world biologicals enterprise. With such background, I give a brief and selective overview of regulatory control, illustrating important points, mainly with viral products. The need to present the flow of history, with brevity, prompted me to rely heavily on personal experiences with limited references.

Before the League of Nations

Biological standardization and control (Table 1) were born of necessity following the late 19th century breakthroughs in microbiology, immunology, virology and vaccinology. These early advances were made by illustrious pioneers including Koch, Pasteur, von Behring and Ehrlich who followed in the footsteps of Edward Jenner [3]. Jenner was the late 18th century initiator of the science of vaccines itself [4]. Development of biologicals standards were evolved as technological advances created increasing needs. The original intent for biologicals control was to establish standard preparations from which units of activity that quantify potency could be derived, but it has proceeded far beyond that initial purpose.

League of Nations [see refs. 5-9]

Although biological standardization was pioneered in a number of countries including Denmark, France, Germany, the United Kingdom and the USA, it did not achieve wide international status until after World War I following the creation of the League of Nations in 1924 and its formation of the Permanent Commission on Biological Standardization of the League's Health Organization [7]. The National Institute for Medical Research in Hamstead, U.K., and the State Serum Institute in Copenhagen became the first custodians of the biological standards that defined potency and safety. This activity was expanded to larger numbers of nations following the Intergovernmental Conference of 1935 [7]. Importantly, it was this conference which recommended that each country should establish its own national regulatory control authority.

Table 1: Evolution of concepts and development of biologicals regulatory standards and practices, pre-World War II.

1796-1923	An era recognizing need for biological standardization. Logical outgrowth of birth of vaccinology and immunology. Luminaries included Jenner, Ehrlich, Koch, Pasteur, von Behring. Early regulation confined to individual nations, Particularly Denmark, France, Germany, UK, USA
1924	Post-World War I began international regulation under League of Nations and its Permanent Commission on Biological Standardization. Expansion to many nations in 1935. Concepts: Each nation to have a national regulatory control Flawed concept: control of biologicals by examination of the end product. Eclipse in 1939 with start of World War II.

It is of interest that Gautier's [5] report in 1935 on the League of Nations activities did not cover vaccines. Instead, it emphasized biological substances for control of disease, and principal focus was given to antisera against bacteria and their toxins. In addition, it was espoused that biological methods in standardization should be discarded as soon as chemical purity of the material was achieved and its chemical and physical constants had been defined.

At that time, chemistry was the dominant science. It antedated the separation of vaccines and antisera from chemically definable substances such as insulin, thyroxine and vitamins. The complexities of the chemicals comprising biological structure and function were not understood, and knowledge of biological systems that rely on protein folding and structural chemistry were to be decades in the future.

The World Health Organization

International biological regulatory affairs went into eclipse with the onset of World War II (WW II) and the death of the League of Nations. They were reborn in 1948 (Table 2) with the formal establishment of the World Health Organization (WHO) as an agency of the United Nations [7-9]. The WHO created the Biologics unit which acts in conjunction with the Expert Committee on Biological Standardization. Reference materials are held by several nations in Europe and the USA.

The entry of WHO into regulatory affairs was marked by recognition that biological products are usually highly complex and cannot be assayed for safety and efficacy by examining the final material alone. Hence regulations have been evolved that cover characterization of all starting materials together with in-process monitoring of all stages of production from start to finish. With time, standards for acceptability of manufacturing and quality control facilities and for good laboratory and manufacturing practices were promulgated. Most recently, especially since the appearance of AIDS, the WHO has sought to facilitate clinical trials of experimental

Table 2: Evolution of concepts and development of biologicals regulatory standards and practices, post-World War II.

1948	United Nations formally establishes the World Health Organization (WHO). WHO establishes continuity from the League of Nations. WHO establishes Biologics unit that acts in conjunction with the Expert Committee on Biological Standardization. Concepts: Biologicals are complex and need to be monitored at all stages of manufacture. New Product standards and regulations prepared with participating international experts. Established world reference centres. Proactive role in defining quality and practices in manufacturing and in quality control. Increasing pursuit of uniform worldwide regulation and standards. Promote clinical studies in developing nations under defined ethical principles

biologicals in developing nations where attack rates are high, and to define the ethical principles [see 46] that must be followed in their conduct.

In its original concept, WHO biological regulations were to serve only as an advisory to national control authorities of member nations [7] unless the WHO regulations were incorporated into those of the national control authority. With time, WHO has become increasingly engaged in seeking to evolve uniform world standards to which all countries eventually may abide. This may be timely since trivial differences may exist between WHO regulatory standards and those of highly qualified national control authorities. The result may be unfair restraint of trade if purchase is restricted exclusively to products meeting WHO standards.

Sources for Possible Adverse Reactions to Viral Vaccines

Adverse reactions to vaccines (Table 3) were not uncommon in the past and have been chronicled in the literature [10, 11]. It is important that the numbers and kinds of adverse events have grown progressively fewer with time and technical sophistication. Past and possible future adverse reactions may focus mainly on the virus itself, on retained or added excipients, or on largely idiosyncratic immunological responses. Past problems with virus vaccines resided mainly in the quality of live virus strains or with the completeness of inactivation of killed virus preparations.

Historic Examples

A review of past problems with individual vaccines (Table 4) [10, 11] provide a window to view what might occur in the future. The oldest of the vaccines, those

Table 3: Possible sources of adverse clinical experiences with vaccines that require special attention.

Vaccine virus itself		
	Live	Suboptimal balance of immunogenicity with reactogenicity. Excess reactogenicity in the immunocompromised person. Reversion to virulence, especially if transmissible.
	Killed	Incomplete inactivation.
Excipients		
	Substrate (cultivation) – retained residua; microbes; media components; DNA. Added stabilizers; adjuvants; antibiotics; preservatives. Diverse inflammatory, carcinogenic, or sensitizing substances.	
Immunological responses		
	Immunopotentiation on subsequent exposure to antigens of complementary specificity. Induction of autoimmune or immune complex disorders. Preordination of inappropriate or imbalanced cellular:humoral immune responses on memory recall.	

Table 4: Historic examples of past problems with viral vaccines.

Vaccine	Experience
Vaccinia	Progressive vaccinia; eczema vaccinatum; post-vaccinal neuropathy.
Rabies	Incomplete inactivation; neural demyelination.
Yellow fever	Retained neurotropism; stabilization with serum from hepatitis B carriers contaminated with the virus.
Influenza	Guillain-Barre polyneuritis syndrome after swine influenza vaccine.
Poliomyelitis	
Killed	Incomplete inactivation of poliovirus; indigenous monkey polyoma SV_{40} virus contamination.
Live	Retention of or reversion to neurovirulence; SV_{40} virus contamination.
Measles	
Live	Insufficient attenuation; immunodeficiency (early life, excessive titer); needed development of leukosis-free chickens.
Mumps – Live	Most mumps vaccine strains other than the Jeryl Lynn virus are excessively neurotropic.
Rubella – Live	Transient arthritis in some adults.

against smallpox and rabies, gave the largest difficulties and were caused by incomplete inactivation, progressive and extensive infections, or with demyelinative neuropathy. Attenuated yellow fever vaccine retained neurotropism or caused hepatitis B when contaminated by human serum used to stabilize the virus.

Rare occurrence of Guillain-Barre polyneuritis following administration of swine influenza vaccine in 1976 in the USA [12, 13] was probably real but was blown to large proportions by a mass immunization programme. The programme itself was the result of misjudgment that a few human cases of influenza which followed introduction of zoonotic swine influenza into a military post represented a true antigenic shift and was the forerunner of a worldwide pandemic such as occurred in 1918-1919 [14]. The virus did not spread beyond the camp and its legacy was an explosive occurrence of Guillain-Barre syndrome which has not been seen again since that time.

Live and killed poliovirus vaccines were the first to utilize virus grown in cell culture (*Macacus* monkey renal cells) and caused the best known [see 11, 15] vaccine problems of this century. The virus in some early killed poliovaccines was not completely inactivated, and live virus vaccines were not sufficiently attenuated. The latter caused iatrogenic poliomyelitis due to retained neurovirulence or to reversion to virulence on passage in humans. A small number of such cases still occur each year. Both live and killed vaccines were plagued in their early period by contamination with undetected indigenous *Macacus* monkey polyomavirus, SV_{40}, [11, 15-18] that was present in the renal cell cultures. Substitution of monkey kidney source using that of *Cercopithecus* monkeys provided both the means to detect the presence of SV_{40} virus and to eliminate it. No causal relationship between SV_{40} virus in vaccine and human disease has been established.

The single live measles, mumps, and rubella, and combined MMR vaccines caused few problems [11, 16, 19, 20] but fine-tuning of the virulence of measles [11, 21] and rubella vaccines [11, 22] was needed. Duck embryonic cell cultures [22] for propagating and attenuating rubella virus was developed and avian leukosis virus was eliminated from chick cell cultures by using embryos from eggs from specific pathogen-free chickens [11, 23]. Most mumps virus strains used to prepare live mumps vaccine retain neurotropism but such problem does not exist with the Jeryl Lynn strain [11, 16, 19, 20].

Vaccine Technologies, Triumphs, and Troubles since 1950

The past 50 years of vaccinology (Table 5) have been seminal and have changed the face of the world through immunoprophylaxis. Smallpox virus was eradicated by intensive vaccination by 1974 [24] and has brought huge societal benefits and annual economic savings. Poliovirus and measles virus eradication are being pursued at this time [16, 25].

Poliovirus vaccine initiated the cell culture era and the discovery of contaminating SV_{40} [11, 17, 18] interrupted its application for a brief period of time. Use of specific pathogen-free animal tissue sources largely eliminated the problem of indigenous viral contamination of primary cell cultures [17, 18, 23]. The more sophisticated and positive solution to indigenous virus contamination was provided

Table 5: Selected events that illustrate the flow of technologies, troubles and triumphs or vaccines in the past five decades.

• Eradication of smallpox.
• Development of cell culture for growing viruses.
• Development of live and killed poliovaccines.
• Indigenous contaminating viruses, discovery of SV_{40}.
• Development of measles, mumps, rubella and combined MMR vaccine. Elimination of indigenous avian leukemia virus. Appropriate attenuation. Duck cell culture (rubella). Efficacy and safety in combined formulations.
• Swine influenza fiasco.
• Cell propagation of hepatitis A virus.
• First licensed anti-cancer virus vaccines. Marek's disease (avian neurolymphomatosis). Plasma-derived hepatitis B vaccine (human hepatocarcinoma). New era of recombinant expression – hepatitis B vaccine.
• Experimental Alexander hepatocarcinoma cell culture for hepatitis B vaccine (development of principles for use of cancer cells).

by Hayflick's [26-28] development of serially passaged human cell cultures that were devoid of markers for neoplasia and were safe for preparing vaccines for man. Though the diploid WI-38 and MRC-5 human cell strains have been used routinely for many years, their original acceptance was the subject of a battle of many years' duration [26-29] in which Dr. Leonard Hayflick and the late Dr. Frank Perkins of WHO Biologics, representing new era science, were engaged in argumentative combat with an old era establishment that was propelled by an arrogant confidence derived from overwhelming ignorance. Hayflick suffered professionally in his refusal to yield [29]. He and Perkins need to be remembered as two of the largely unsung heroes to whom preventive medicine by vaccines owes so much. The swine influenza vaccine fiasco [12, 13] serves, to this day, as a cogent reminder that baseless pursuit of an overwhelmingly improbable event is injudicious and should never be repeated.

Solution of problems in the original development of the measles, mumps, and rubella vaccines in our laboratories [16, 19-22] gave way to the pioneering development of combined triple live vaccine MMR. Though developed 27 years ago, MMR remains a principal flagship for paediatric immunization in the developed nations and is now being introduced worldwide. The discovery of hepatitis A virus in 1973 [16, 30] and its breakthrough propagation in cell culture in 1979 [16, 31] opened the way to both live and killed virus vaccines [16, 32].

Many cancers of animals and about 15% of cancers of human beings are caused by viruses. Viral vaccines may be very important in prevention [33]. First successes in developing licensed vaccines against animal and human cancer were accomplished in our laboratories [34-39]. Live wet and dried vaccines against neurolymphomatosis of chickens were developed using Burmester's turkey herpesvirus [34, 35] and were licensed in 1971 and 1975. Hepatitis B virus, a principal cause of human cancer, was an enigma until the discovery of viral surface antigen in the plasma of human carriers of the infection [see 16, 36-39] that opened the door to development of the world's first licensed vaccine against cancer in man and that is now widely used in most nations of the world. Development of recombinant yeast-derived hepatitis B vaccine [40-42] was the first genetically engineered vaccine of any kind.

Development and testing of an experimental hepatitis B vaccine derived from human hepatocarcinoma (Alexander) cells in culture that shed surface antigen, was the first example [43-45] of how safe and effective vaccines can be derived from antigens produced using serially propagable human cancer cells by high-level purification and DNA removal established and monitored through validated manufacturing and control processes.

Regulatory Facilitation of New Product Development

The concepts and procedures (Table 6) that have been evolved and promulgated over the years are extremely helpful in facilitating new product development and production and in providing reliable guidelines for their early basic explorations. Guidelines and points to consider became very important with the new era of genetic engineering. Partnering of industry with regulatory agencies that preserve confidentiality is of immense help as a reliable resource which limits injudicious and frivolous pursuits. If the regulatory establishment did not already exist, it would surely need to be invented.

Table 6: Examples of concepts and procedures of regulatory agencies that facilitate viral vaccine development and production.

High level purity, control of incipients.
Optimal balance in toxicity vs. efficacy.
Frozen seed lot systems for viruses and cells.
Freedom from contaminants from animal tissues.
Fail-safe systems, redundancy, and validation in in-process monitoring.
Guidelines and points to consider in product development, e.g. Recombinant vector expression. Recombinant DNA plasmid vaccines.
Partnering with industry in R&D while protecting confidentiality.
Regulatory definition of standards and promulgation of regulations to assure stability, safety and product efficacy.
Product licensure.

Reasons to Like and to Appreciate Regulators

Regulators sometimes have been considered by novice vaccinologists to be a drag on progress. These initiates soon learn otherwise. The necessary attributes and contributions of regulators are easily definable (Table 7). In overview, they consist mostly of intelligent and skilled workers of widely diverse knowledge and experience who serve importantly and unselfishly in the enterprise for improving human health through development and use of vaccines. Exercise of judgment and compromise, and participation in logical, legal, and ethical [46] issues are necessary and routine engagements. Those of us who labour in the more visible aspects of vaccinology sometimes fail, I believe, to give sufficient credit and appreciation to the regulators upon whom we rely for so much.

Table 7: Some reasons to like and to appreciate regulators!

Complex body of Renaissance People who see the big picture.
Represent great depth and diversity in science and technology.
Display infinite patience while creating the whole from overwhelming minutiae.
Exercise judgment and create reasonable compromise.
Employ lawyer logic and deal with legal issues.
Under Investigative New Drug Laws and requirements for good clinical practices, engage in current ethical standards of autonomy, beneficence and justice to protect the rights of volunteer participants in clinical studies.
(They can be a bit nosey and stuffy!)

Closing

I would close with two cartoons that sum up my view of the historic relationships between the regulators and the regulated. In the early days (Fig. 1), the regulators tended to stand aloof and separated from the regulated by an arm's length. This may have been a means to maintain fairness, to remain free of bias that might result from overt friendship, or to cover for lack of information. That has all been changed (Fig. 2). The regulators and the regulated now express their mutual respect and esteem in interactive nose-to-nose engagements.

Fig. 1: LIKE IT WAS. Arms Length – Rigid Regulation of the Regulated.

Fig. 2: LIKE IT IS. Interactive Regulation of the Regulated.

Acknowledgement

We are indebted to Sandra Hobbis for the artistic creativity and for preparing the cartoons.

References

1 Anonymous: 50 years of making people healthier. World Health Organization Magazine. 51st year, issue 2, 1998, 31 pages.

2 Hussein A: WHO at fifty. Highlights of the early years until 1960. World Health Forum 1998;19: 21-37.

3 Plotkin SB, Plotkin SA: A short history of vaccination, in Plotkin SA, Mortimer EA Jr (eds): Vaccine. Philadelphia, WB Saunders Co, 1988, ch 1, pp 1-7.

4 Jenner E: An Inquiry into the Causes and Effects of the Variolae Vaccinae. London, Low, 1798.

5 Gautier R: The health organisation and biological standardisation. Quarterly Bulletin of the Health Organisation, League of Nations. 1935;IV:497-554.

6 Gautier R: The health organisation and biological standardisation (second memorandum). Bulletin of the Health Organisation, League of Nations. 1945/46;XI:1-75.

7 Outschoorn AS: The biological standardization programme of the World Health Organization. J Biol Stand 1973;1:203-213.

8 Sizaret P: Evolution of activities in international biological standardization since the early days of the Health Organisation of the League of Nations. Bull WHO 1988;66:1-6.

9 Cockburn WC, Hobson B, Lightbown JW, Lyng J, Magrath D: The international contribution to the standardization of biological substances. III. Biological standardization and the World Health Organization 1947-1990. Specific activities and commentary. Biologicals 1992;20:1-10.

10 Wilson Sir GS: The Hazards of Immunization. University of London, The Athlone Press, 1967, 324 pages.

11 Hilleman MR: Viral vaccines in historical perspective, in Non Target Effects of Live Vaccines. Dev Biol Stand. Basel, Karger, 1995, vol 84, pp 107-116.

12 Silverstein AM: Pure Politics and Impure Science: The Swine Flu Affair. Baltimore, Johns Hopkins University Press, 1981.

13 Hilleman MR: Cooperation between government and industry in combating a perceived emerging pandemic. The 1976 swine influenza vaccination program. J Amer Med Assoc 1996;275: 241-243.

14 Webster RG, Kawaoka Y: Influenza: an emerging and reemerging disease. Semin Virol 1994;5: 103-111.

15 Robbins FC: Polio-historical, in Plotkin SA, Mortimer EA Jr (eds): Vaccine. Philadelphia, WB Saunders, 1988, ch 6, pp 98-114.

16 Hilleman MR: Six decades of vaccine development – a personal history. Nature Medicine Vaccine Supplement 1998;4:507-514.

17 Sweet BH, Hilleman MR: The vacuolating virus, S.V. 40. Proc Soc Exper Biol Med 1960;105: 420-427.

18 Hilleman MR: Discovery of simian virus 40 (SV_{40}) and its relationship to poliomyelitis virus vaccines, in Brown F, Lewis AM Jr (eds): Simian Virus 40 (SV40): A Possible Human Polyomavirus. Dev Biol Stand. Basel, Karger, 1998, vol 94, pp 183-190.

19 Hilleman MR: The development of live attenuated mumps virus vaccine in historic perspective and its role in the evolution of combined M-M-R, in Plotkin S, Fantini B (eds): Vaccinia, Vaccination and Vaccinology: Jenner, Pasteur and their successors. Paris, Elsevier, 1996, pp 283-292.

20 Hilleman MR: Combined measles, mumps, and rubella vaccines, in Ellis RW (ed): Combination Vaccines. Totowa NJ, Humana Press (in press).

21 Hilleman MR, Buynak EB, Weibel RE, Stokes J Jr, Whitman JE Jr, Leagus MB: Development and evaluation of the Moraten measles virus vaccine. J Amer Med Assoc 1968;206:587-590.

22 Hilleman MR, Buynak EB, Whitman JE Jr, Weibel RE, Stokes J Jr: Live attenuated rubella virus vaccines. Experiences with duck embryo cell preparations. Amer J Dis Child 1969;118: 166-171.

23 Hughes WF, Watanabe DH, Rubin H: The development of a chicken flock apparently free of leukosis virus. Avian Dis 1963;7:154-165.

24 Fenner F, Henderson DA, Arita I, Ladnyi ID: Smallpox and its eradication. Geneva, World Health Organization, 1988.

25 de Quadros CA: Global eradication of poliomyelitis and measles: Another quiet revolution. Ann Intern Med 1997;127:156-158.

26 Hayflick L: History of cell substrates used for human biologicals, in Hayflick L, Hennessen W (eds): Continuous Cell Lines as Substrates for Biologicals. Dev Biol Stand. Basel, Karger, 1989, vol 70, pp 11-26.

27 Hilleman MR: History, precedent, and progress in the development of mammalian cell culture systems for preparing vaccines: Safety considerations revisited. J Med Virol 1990;31:5-12.

28 Hilleman MR: Line cell saga – An argument in favor of production of biologicals in cancer cells, in Petricciani JC, Hopps HE, Chapple PJ (eds): Cell Substrates. New York, Plenum Publishers, 1979, pp 47-58.

29 Hayflick L: A novel technique for transforming the theft of mortal human cells into praiseworthy federal policy. Experimental Gerontology 1998;33:191-207.

30 Provost PJ, Ittensohn OL, Villarejos VM, Arguedas G JA, Hilleman MR: Etiologic relationship of marmoset propagated CR326 hepatitis A virus to hepatitis in man. Proc Soc Exper Biol Med 1973;142:1257-1267.

31 Provost PJ, Hilleman MR: Propagation of human hepatitis A virus in cell culture in vitro. Proc Soc Exper Biol Med 1979;160:213-221.

32 Hilleman MR: Hepatitis and hepatitis a vaccine: A glimpse of history. J Hepatology 1993;18: S5-S10.

33 Hilleman MR: Overview of viruses, cancer and vaccines in concept and in reality. Recent Results in Cancer Research 1998;154:345-362.

34 Hilleman MR: Marek's disease vaccine. Its implication in biology and medicine. Avian Dis 1972;16:191-199.

35 Okazaki W, Purchase HG, Burmester BR: Protection against Marek's disease by vaccination with a herpesvirus of turkeys. Avian Dis 1970;14:413-429.

36 Buynak EB, Roehm RR, Tytell AA, Bertland AU, Lampson GP, Hilleman MR: Vaccine against human hepatitis B. J Amer Med Assoc 1976:235:2832-2834.

37 Hilleman MR: Plasma-derived hepatitis B vaccine – A breakthrough in preventive medicine, in Ellis RW (ed): Hepatitis B Vaccines in Clinical Practice. New York, Marcel Dekker, 1979, ch 2, pp 17-39.

38 Hilleman MR: Immunology, vaccinology and pathogenesis of hepatitis B, in Koprowski H, Oldstone MBA (eds): Microbe Hunters, Then and Now. Bloomington IL, Medi-Ed Press, 1996, ch 16, pp 221-233.

39 Hilleman MR: Three decades of hepatitis vaccinology in historic perspective. A paradigm of successful pursuits, in Plotkin S, Fantini B (eds): Vaccinia, Vaccination and Vaccinology: Jenner, Pasteur and Their Successors. Paris, Elsevier, 1996, pp 199-209.

40 McAleer WJ, Buynak EB, Maigetter RZ, Wampler DE, Miller WJ, Hilleman MR: Human hepatitis B from recombinant yeast. Nature 1984;307:178-180.

41 Hilleman MR: Yeast recombinant hepatitis B vaccine. Infection 1987;15:2-6.

42 Hilleman MR, Weibel RE, Scolnick EM: Recombinant yeast human hepatitis B vaccine. J Hong Kong Med Assoc 1985;37:75-85.

43 McAleer WJ, Markus HZ, Bailey FJ, Herman AC, Harder BJ, Wampler DE, Miller WJ, Keller PM, Buynak EB, Hilleman MR: Production of purified hepatitis B surface antigen from Alexander hepatoma cells grown in artificial capillary units. J Virol Meth 1983;7:263-271.

44 McAleer WJ, Markus HZ, Wampler DE, Buynak EB, Miller WJ, Weibel RE, McLean AA, Hilleman MR: Vaccine against human hepatitis B virus prepared from antigen derived from human hepatoma cells in culture. Proc Soc Exper Biol Med 1984;175:314-319.

45 Hilleman MR, Buynak EB, Markus HZ, Maigetter RZ, McAleer WJ, McLean AA, Miller WJ, Wampler DE, Weibel RE: Control of hepatitis B virus infection: vaccines produced from Alexander cell line and from recombinant yeast cell cultures, in Vyas GN, Dienstag J, Hoofnagle J (eds): Viral Hepatitis and Liver Disease. Orlando FL, Grune & Stratton, 1984, pp 307-314.

46 Hilleman MR: Vaccines, human experimentation, and ethics in evolutionary perspective, in Brown F, Plotkin S (eds): Preclinical and Clinical Development of New Vaccines. Dev Biol Stand. Basel, Karger, 1998, vol 95, pp 13-17.

Dr. M.R. Hilleman, Merck Institute for Therapeutic Research, Merck Research Laboratories (WP53C-350), 770 Sumneytown Pike, West Point, PA 19486, USA.

Brown F, Griffiths E, Horaud F, Schild GC (eds): A Celebration of 50 Years of Progress in Biological Standardization and Control at WHO.
Dev Biol Stand. Basel, Karger, 1999, vol 100, pp 31-37

Standardization of Smallpox Vaccines and the Eradication Programme – a WHO Perspective

I. Arita

Agency for Cooperation in International Health (ACIH), Kumamoto, Japan

Key words: Smallpox vaccine, smallpox eradication, international testing service.

Abstract: Smallpox vaccine was born in 1796 and the need for it ended in 1980 with the eradication of smallpox. What was remarkable was that in its 200 years history, when the global smallpox eradication programme really needed the vaccine in 1967, it was found that, on the whole, quality was unsatisfactory and supply was short. Unconventional counter measures were taken by WHO, including an international independent testing system, provision of reference preparations in large quantity etc. They effectively solved the problems meeting the target eradication schedule of 10 years. The lessons learned would be of some value to the future eradication or effective control programme of vaccine preventable diseases.

A history of smallpox vaccine is reviewed in this article. Special attention is paid to the problems of smallpox vaccine when the global eradication of smallpox was carried out from 1967 to 1980. The findings may be of interest to those concerned with control of vaccine preventable diseases.

Development of Smallpox Vaccine Product

In late 18th century, the vaccine started with cow-pox virus in Sarah Nelme's hand (Fig. 1). In 19th century, as it was a pioneer of a new product, it was not surprising that the product had to be used, first, person to person, then animal skin to person: and finally calf-lymph in ampoules or vials. During the first half of the 20th century, the liquid type of calf lymph was a common production method.

The invention of freeze-dried vaccine in the 1950s further facilitated the wide use of the vaccine in less developed regions. Vaccination programmes and outbreak response in many countries resulted in circumscribing smallpox endemic areas to developing countries of Latin America, the Asian continent, Indonesia and Africa.

The period of the 1960s was specifically active for the refinement of smallpox vaccine products in terms of study on vaccine complications done in the US, and the search for less pathogenic vaccinia strains and tissue culture vaccine in Europe and North America. Collaborating studies organized by WHO led to the establishment of

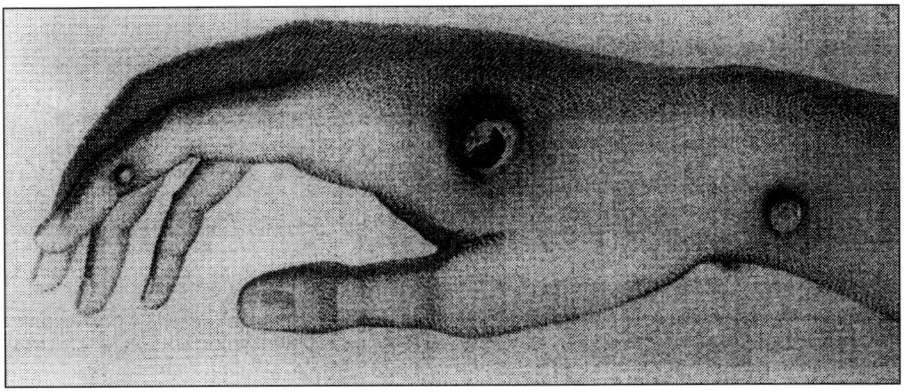

Fig. 1: Accidental cowpox lesions on the hand of Sarah Nelmes (case XVI in Jenner's Inquiry from which material was taken for the vaccination of James Phipps in 1796.

Minimum Requirements for smallpox vaccine and an international standard preparation of smallpox vaccine.

Smallpox vaccine and Intensified Smallpox Eradication Programme

The global smallpox eradication started in 1958 with the WHA (World Health Assembly) resolution, but the progress was slow. In 1967, with new WHA resolution, an intensified programme was initiated, resulting in development of national eradication programmes in 31 smallpox endemic countries. WHO took responsibility for coordinating the implementation of this global programme.

Because a sufficient supply of heat stable quality vaccines to these national programmes was prerequisite to success,WHO surveyed the real situation of the vaccine quality and supply throughout the world [1]. Smallpox freeze-dried vaccine is termed as the vaccine in the following sections.

The WHO 1967 survey revealed that: (i) throughout the world, there were 77 laboratories in 52 countries producing smallpox vaccines; (ii) of 77 laboratories, 59 reported their results of vaccine testing (Table 1). For potency; 47% were unsatisfactory and for heat-stability, 87% unsatisfactory. Further, an independent testing of the vaccine proposed for donation or bilateral contribution resulted in unsatisfactory results (Table 2). (iii) In bacterial counts, the results were better; only 8% failed to meet WHO requirements. On the whole, it was strongly suspected that sub-standard vaccines were being used in most national eradication programmes. In particular, unsatisfactory results of heat stability gave considerable concern, since 31 endemic countries were situated in tropical areas.

Counter Measures

Over the previous 150 years, a large number of laboratories started to produce vaccines in both developed and developing countries. The requirements such as personnel,

Table 1: WHO survey.

Laboratory		Initial potency		Heat stability	
		Satisfactory	Unsatisfactory*	Satisfactory	Unsatisfactory*
Africa	8	4	4	2	6
Americas	13	6	7	2	11
Asia and Oceania	16	9	7	5	11
Europe	22	12	10	7	15
Total	59	31	28 (47%)	16	43 (87%)

* Lack of consistency, unreliable tests and no report included. Not only labs in Africa, a few manufactures in industrialized countries failed meeting WHO standards.

Table 2: Independent testing in 1967 by the WHO international reference centre for smallpox vaccine in Bilthoven of production batches intended for use in the eradication programme.

Continent	Proposed for use in the eradication programme through donation to WHO or through bilateral contributions		
	Number of producers	Number of batches tested	Number of batches unsatisfactory (%)
Americas	1	1	1
Africa	1	1	0
Asia	2	11	9
Europe	2	22	12
Total	6	35	22 (62)

plant, production method, tests during the production process and of final product and control practice of national authorities varied tremendously according to individual laboratories. Some standardization effort such as collaborating studies was exercised in the past, but it became obvious that it had no effect. In the circumstances, the following counter measures were taken under the strong leadership of WHO:

- Two WHO International Reference Centres for Smallpox Vaccine (termed as WHO Vaccine Centre in the following sections) were designated in 1967. The first was the National Institute of Public Health, Bilthoven, the Netherlands and the second, Connaught Medical Research Laboratories, University of Toronto, Canada, as Regional Centre. These acted as the technical centres for collaborating with WHO for the implementation of counter measures.

- In these Centres, international testing services were carried out. All laboratories producing the vaccines for donation or for their own national eradication programmes sent their final products as test samples and only the vaccines meeting WHO requirements were used for the programmes.

- The National Institute of Public Health, Bilthoven, produced a large quantity of «working» reference preparations (standardized with International Reference Preparation) and seed virus lots. These preparations were supplied to any laboratories on request.
- In 1968 a special study group was formed to visit two production laboratories in the U.S. and Russia and developed a production manual, which was practical, simple and up to date. These were distributed to all laboratories.
- WHO arranged for production laboratories to receive frequent visits by consultants who advised on the improvement of production and testing procedures.

Improvement of Vaccine Quality

Regularly, 20-30 production laboratories sent samples to two WHO Centres. From 1967 to 1969 a marked improvement was noticed (Table 3). Virtually since

Table 3: WHO quality control of freeze-dries vaccine: results of tests carried out in WHO reference centres for smallpox vaccine in Bilthoven and Toronto on experimental and production batches from producers.

Year	Number of producers	Number of batches submitted	Number satisfactory (%)	
1967	20	74	27	(36)
1968	23	136	74	(54)
1969	30	164	128	(78)
1970	27	380	312	(82)
1971	32	206	154	(75)
1972	27	311	241	(77)
1973	30	392	367	(94)
1974	28	231	199	(86)
1975	21	167	139	(83)
1976	16	213	203	(95)
1977	11	114	101	(89)
1978	9	59	57	(97)
1979	10	85	82	(96)
1980	5	46	46	(100)
Total	–	2,578	2,130	(82.6)

1970, all the vaccine batches meeting WHO requirements were used for the eradication programme (either by donation or domestic production). The provision of «working» reference preparations and seed lots with Lister strain was most helpful since many laboratories were unable to produce working preparations, although International Reference Preparations were intended to serve that purpose.

At the beginning of testing services, some national health authorities complained that the system was a censorship-type procedure. However, when they realized that services were for their own benefit, such as technical advice, provision of reference preparations, seed lots ampoules etc., considerable cooperation had developed.

The 1967 WHO survey revealed that vaccinia virus strains varied widely according to laboratory [1]. WHO informally recommended the New York Board of Health and Lister strains, due to their lower pathogenicity and good immunogenicity. In fact, in 1968, individual laboratories in 49 countries could be divided into a group of 28 laboratories using the New York Board of Health or Lister strains and another group of 33 using different strains. In 1971, laboratories using either of these two strains increased to 47.

The further attenuated vaccinia strains were studied by many laboratory workers in the 1960s and 1970s [2]. Late in the 1970s, LC16m8 (Japan) strain was developed as a workable attenuated strain, due to its weak pathogenicity as well as good immunogenicity, but it was not used for the global programme since it was too late [3]. A tissue culture vaccine with RK cells (the Netherlands) was also developed around this time, but again it was too late for actual use.

Constant Flow of the Quality Vaccine

In 1967, WHO estimated that 300 million doses of vaccine would be required every year for the global programme, namely, the vaccine supply to smallpox-endemic countries and their adjacent countries. The WHO budget was short and donations, although intensified since 1967, varied yearly from 15 to 50 million doses only. Thus there was serious concern about the shortage of the vaccine supply. The problems were solved with the following counter measures:

- WHO testing services rapidly increased the number of batches acceptable for release in the field (see Table 3).
- The WHO survey revealed that many laboratories were producing 50 to 100 dose ampoules or vials which resulted in a substantial wastage of the vaccine in the field where house-to-house vaccination was in principle recommended. Guidance was given to all laboratories to produce 10 to 20 dose ampoules or vials, which substantially reduced the wastage.
- Bifurcated needles were introduced with Wyeth Laboratories' agreement to exempt the patent. This method permitted four times as many vaccinations to be given by the conventional method. (Bifurcated needles – 0.0025 ml of reconstituted vaccine needed, conventional method 0.01 ml.)
- In the 1970s, all national eradication programmes were using the bifurcated needles. The small dose of 0.15-0.25 ml (10 to 20 dose) ampoules or vials were also suited to this method. It should be noted that the use of bifurcated needles was

simple and produced the satisfactory or better take rates. Special containers for sterilization were also developed.

- Major endemic areas of Bangladesh, India and Indonesia required altogether 100 million-dose vaccine yearly. As the programmes operated in full swing, the requirements increased yearly up to 150 million doses by the mid 1970s. WHO, in collaboration with WHO Vaccine Centres, made substantial efforts by testing services as well as dispatching consultants to improve the vaccine quality and at the same time to augment the production capacities. In fact, the supply requirements as mentioned above were met by their own domestic production in these three countries.
- Countries of low population (less than 10 million) were not advised by WHO to encourage local production which would not be economical. The exception was Guinea, which produced vaccine for a limited period.
- In 1980, WHA announced the success of the global eradication of smallpox and urged the member states to implement the immediate discontinuation of smallpox vaccination, as recommended by the Global Commission for Certification of Smallpox Eradication. Furthermore, to assure safety, until 1986, WHO kept a vaccine reserve (200 million doses) which could be supplied to anywhere in the world in case of emergency. The discontinuation of smallpox vaccination progressed and today there is no known national programmes of smallpox vaccination in the world.

Conclusion and Lessons

The vaccine, in its final years of life, was most needed to make the global smallpox eradication programme succeed. However, it was found that the vaccine was of extremely unsatisfactory quality and in short supply to meet such requirements. Control measures to solve the difficulties were unconventional, but achieved the task effectively.

Perhaps some further discussions on these experiences would be useful, when we think of emerging vaccines as well as future programme for effective control or eradication programmes for vaccine-preventable diseases.

WHO testing service handled only the final products. The system identified a large proportion of production batches as unsatisfactory. This compelled national health authorities as well as WHO to act. The principle of biological standardization is that vaccine quality control or assessment cannot be done by just testing «final products». It should cover personnel, facilities, production process, final product, national control etc., namely, all aspects of GMP. If we had followed this «comprehensive» approach, however, we should not have been able to resolve the problems.

In the 1950-1970s, extensive studies were continuing in terms of research into vaccine strains, attenuated strains, tissue culture or egg vaccines etc. WHO did not allocate much funding and time to promote these studies, but concentrated on practical studies such as moisture content of the vaccine, «rapid» heat-stability tests, assessment of laboratory testing errors etc. The limited funds necessitated WHO taking this approach, but the programme successfully met its 10-years target.

The LC16m8 strain could be a most promising strain with which a new recombinant vaccine may be attempted with tissue culture and the use of bifurcated needles. It would also allow the rapid production of a large quantity of the vaccine in case of unexpected emergency circumstances.

The concept of an International Reference Preparation is that the product is precious and, therefore, of limited supply. Usually laboratories receive just two or three ampoules from which a working reference preparation has to be produced by the recipient laboratories themselves. In the smallpox eradication programme, many laboratories were unable to follow these procedures, resulting in unsatisfactory testing. Provision of working preparations by WHO Vaccine Centres solved the problem, an approach applicable for some future vaccine programmes.

Lastly, since one of the important conditions of effective control or eradication programmes of vaccine-preventable diseases is that it is bound to be completed within the time limit of 10-15 years [4], the initiation of any global programme should analyse, before it is started, the feasibility of flow of the quality vaccine, and measures such as these unconventional approaches, described here, may be of some value for the establishment of a practical approach.

Acknowledgement

I am in debt to all the smallpox workers who contributed to this extraordinary vaccine programme during the eradication campaign. Because of their excellent achievements, I was able to write this article. I thank Ms R. Tominaga, ACIH, for her assistance in preparing the article.

References

1 Arita I: The control of vaccine quality in the smallpox eradication programme, Symposia Series in Immunobiological Standardization, Basel, Karger, 1973, vol 19, pp 79-89.

2 Marennikova SS: Evaluation of vaccinia strains by their behaviour in vaccinated animals and possible implications of the revealed features of smallpox vaccine practice, Symposia Series in Immunobiological Standardization, Basel, Karger, 1973, vol 19, pp 253-260.

3 Hashizume S, Yoshizawa H et al: Vaccinia Viruses as Vectors for Vaccine Antigens, Elsevier Science Publishing Co., Inc., 1985, pp 87-99.

4 Arita I: Are there better global mechanisms for formulating, implementing and evaluating eradication programs?, Report of the Dahlem Workshop on the Eradication of Infectious Diseases, Berlin, March 16-22, John Wiley & Sons, 1997, pp 177-186.

Dr. I. Arita, Agency for Cooperation in International Health (ACIH), 4-11-1 Higashi-machi, Kumamoto-city, Kumamoto, 862-0901 Japan.

Brown F, Griffiths E, Horaud F, Schild GC (eds): A Celebration of 50 Years of Progress in Biological Standardization and Control at WHO.
Dev Biol Stand. Basel, Karger, 1999, vol 100, pp 39-47

From Insulin to Amylin: 75 years of Biological Standardisation in Endocrinology

S.L. Jeffcoate

Dorking, Surrey, U.K.

INTRODUCTION

This is an historical survey of the major contribution that biological standardisation has made to the science of endocrinology in the past 75 years. Many attempts have been made by historians to define their discipline, from the cynical and iconoclastic [1] – «ignorance is the first requisite of the historian, ignorance which simplifies and clarifies, which selects and omits» – to the more reflective [2] – «knowledge of the past has come down through one or more human minds, has been processed by them... and since all historical judgements involve persons and point of view, there is no objective historical truth.» History then, like science, is more than an assemblage of facts. It inevitably consists also of interpretation and since the present has no more than a notional existence between the past and the future, it also consists of predictions. These will be guiding thoughts for my review.

THE PRECEDING 25 YEARS; THE LEGACY OF THE LEAGUE OF NATIONS, 1923-1948

In 1923, shortly after the initial extraction of insulin from the mammalian pancreas and the first clinical use in diabetes, it was recognised that patients – and the clinicians upon whom they depended – needed the different manufacturers and academics to develop specific approaches to standardisation to ensure the potency, authenticity and comparability of the different insulin products, both batch-to-batch for each product and on a between-product basis.

Many of these principles and practices had already been discussed and established in other fields, notably in the treatment and prevention of infectious diseases (as described elsewhere in this volume). Thus in 1902 in the United States Code of Federal Regulations (which had been repeated verbatim in the 1919 CFR) *«biologic products»* had been defined, or rather categorised, as «vaccines, sera, toxins, and analogous products». A similar restrictive approach (by which I mean that the listing of classes of product in a legal document restricts the ability to add or remove items from the list as experience accumulates and knowledge progresses) was adopted in Germany and other European nations and eventually in the post-war European Economic Community.

A rather different, on the surface more vague but in principle and practice more scientifically rigorous, approach was adopted in the United Kingdom. In 1919 a Committee was set up to report to the Minister of Health on the procedures for ensuring the safety of «such substances offered for sale to the public as cannot be tested adequately by direct chemical means.» This expression, implying as it does an evolutionary process as chemical analysis develops, was reproduced in the Therapeutic Substances Act in 1925 which was based on the Report of this Committee and again half a century later – with slight modification (*«biological substances* means substances whose purity and potency cannot, in the opinion of the Secretary of State, be adequately tested by chemical means») – in the U.K.'s Biological Standards Act of 1975.

Experience with the biological standardisation of insulin in the early 1920s coincided with these legal developments (and by the time that the 1925 Act was published, insulin was specifically identified in Schedule 3 as requiring biological testing) and the hormone became a paradigm for the theoretical background and practical solutions to the need for standardisation in endocrinology for over half a century. The need for an *International Standard* for insulin in physical form was established at a meeting in Edinburgh in 1923 after a well-judged intervention by Henry Dale (quoted in [3]). This led to an International Collaborative Study involving the laboratories of the manufacturers and of academic groups which were the fore-runners of the post-war control laboratories. The results were discussed at a second meeting of the group in Geneva in 1925 and the First International Standard (IS) of Insulin was established with a defined content in terms of an *International Unit* (I.U.); the data, discussion and conclusions were published under the auspices of the League of Nations Health Organisation in 1926 [4].

These procedures became well established in the field of endocrinology and used for the next 75 years: (a) the identification of a *need* in the protection of human health is followed by (b) the preparation of one or more *candidate International Standards* which (c) are subjected to *International Collaborative Study*; (d) the results are carefully analysed and, with the agreement of the participating laboratories, an IS with an assigned *International Unit* is established; and (e) finally the findings are published so that all relevant parties and organisations can be informed.

A number of key principles emerged during these early years with increasing clarity and as they still underpin an informed understanding of the subject, I shall enunciate them – in rather simple terms:

- there is no clear cut and permanent distinction between *«chemical»* and *«biological»* medicines and biologicals are effectively chemicals *«in transit»* awaiting chemical procedures to become «adequate» to give assurance of their purity and potency;

- the *International Unit* defined by biological assay is a *unit of biological activity* which allows comparison of results of different manufacturers irrespective of the details of the method used and of the chemical purity of the materials tested;

- as successive *International Standards* replace each other comparability and continuity from one to the next is maintained by assaying them directly against each other in a collaborative assay;
- an IS does not need to be *«pure»* in either of two senses – freedom from impurities or consisting of only one active principle – and the IS gives the basis for continuity of the IU as developments in manufacturing inevitably lead to preparations of increased purity;
- a hormone has two separate (though connected) features – *«structure»* and *«function»* which can each be measured with validity but *not at the same time*. This is reminiscent of Heisenberg's famous Uncertainty Principle with reference to the electron. This entity has the two properties of «position» and «speed» which can be separately measured with accuracy – but not at the same time;
- it is *valid* to measure a *mixture of different structures* – provided the analytical specificity of the method used (whether functionally-directed or structurally-directed) is identical for the different components in the mixture;

Experience with insulin accumulated over 75 years established this hormone as a reference point for these principles and confirmed the basis for biological standardisation of an increasingly wide range of hormones in endocrinology from the 1920s:

- insulin is now a chemically defined drug and there is no longer a requirement for a biological assay to ensure its purity and potency for clinical use;
- the First IS was established for use in a range of bioassays (using frogs, rabbits etc.) and fulfilled this function despite being only about 30% pure (8 IU per mg);
- the Second IS for Insulin was established in 1935 [5] by direct assay against the First IS and maintained continuity of the IU despite being almost three times more pure (22 IU per mg);
- the current, Fifth, IS approaches 100% pure human insulin whereas all previous preparations were less pure material of animal origin. The First IS for example when subjected to the modern technique of HPLC can be shown to consist of similar amounts of bovine and porcine insulin and significant proportions of their de-amidated forms, i.e. at least four «insulins» with identical biological activity;
- this range of active ingredients in the First IS did not prevent its valid and effective use in the bioassays then current because these methods did not distinguish between the different structural forms of the insulin molecules in the clinical preparations;
- the need to differentiate, both in thinking and in practice, between «functionally-directed» and «structurally-directed» assays arises from the above – but has been a more modern concern as discussed below.

In the wider field of endocrinology, other great advances were being made in the years prior to the establishment of WHO. The hormones regulating all the fundamental physiological processes – of growth, reproduction and fertility, nutrition and metabolism, salt and water balance, bones, metabolic rate, and so on – were being identified, assays developed and clinical applications explored. Problems of standardisation were inevitably thrown up and I shall give only one anecdotal example, that of follicle-stimulating hormone, FSH, the complexity of which still causes angst and uncertainty in the late 1990s.

Though FSH is now known to be a glycoprotein from the anterior pituitary gland, in 1931 this was not known and there was at that time confusion between FSH and other closely related gonadotrophins, and even uncertainty between the placental and pituitary sources of the hormone(s). So when my father wrote in his MD thesis in the University of Liverpool [6] of «the follicle-stimulating hormone» he was making a functional statement. A further statement in his thesis is illuminating of the times: «*Although many investigations have been carried out in connection with the standardisation of these hormones yet quantitative work is surrounded with difficulties.* »

The Following 25 Years: World Health Organisation and the ECBS, 1948-1973

«Biological Standardization is a heritage from the Health Organization of the League of Nations, under whose auspices international standards and units have been established for thirty-nine substances which are assayed by biological methods.»

In this way a paper at the First World Health Assembly in 1948 [7] described WHO's inheritance and it went on to describe the objectives of providing ISs for «Therapeutic, Prophylactic and Diagnostic Agents» and the structures need to implement the programme: a *section within the secretariat*; an *expert advisory committee* of not more than ten members – a number always exceeded in my experience of the Expert Committee on Biological Standardization (ECBS) during the period 1986-1993; and «*sub-committees* on antibiotics, antigens, blood-groups, vitamins, hormones, and others as circumstances shall dictate». This last provision has also rarely been implemented in endocrinology despite its crucial importance. In my experience there were a number of occasions at the ECBS when I felt that the multi-disciplinary group would have benefited (and in one instance concerning FSH may have reached a different, more sustainable, decision) if they had received informed advice from a specialist sub-committee.

For hormone therapeutics like insulin (and a range of other biologicals) the required regulatory framework had become clear and, to an extent, was in place. It was accepted that end *Product* testing by bioassays was only one of the approaches needed to assure the quality of such medicines; a guarantee of the quality of four other 'P's was needed – *Premises, Processes, Personnel, Procedures* – all of which were to be examined as part of the licensing and other regulatory processes undertaken by national bodies with, in many instances, expert advice from the ECBS.

In regard to the fifth P, end product testing, there was during this 25 year period a marked advance in the application of chemical analysis to the elucidation of hormone structure and of chemical synthesis to the provision of increasingly pure products for therapeutic use in endocrinology. Chemistry came to take an increasingly dominant position so that in 1973, the end of the period under discussion, the then head of the Biologicals programme at WHO was able to write [8]:

«Biological Standardization is clearly a self-limiting programme. As more is known of a particular biological substance and it can be adequately characterised by chemical and physical means, the need for a biological standard would be eliminated. One may imagine therefore, that the WHO programme of Biological Standardization would get smaller as the years go by.»

Although this may have happened with some biological products such as steroids, vitamins and antibiotics, for protein hormones the programme had never been busier. This was because the chemical advances, striking though they were, needed to be based on a strong foundation of functional understanding and of biological assay of hormones to provide an index of «authenticity» (at least for a while). During this period, progress in the application and usefulness of biological assays was greatly facilitated by advances in statistical design and analysis and the publication by DJ Finney of his landmark text *Statistical method in Biological Assay* in 1952 [9] was a major milestone.

For polypeptides up to about 30 residues (about 4 kDa) it became possible to synthesise analogues for potential clinical use though this was limited by a number of factors including the relative inefficiency and imprecision (by more recent standards) of the synthetic and purification procedures and the analytical techniques used to assess purity. For the larger peptide hormones, and particularly for glycoprotein hormones, for which chemistry had not yet progressed as far as to give adequate information on quality, biological assays and biological standards were still the bed-rock of progress throughout this period.

For a glycoprotein hormone like FSH, progress in unravelling its complex structure was only possible with the tool of a specific, reliable bioassay to hand. For FSH the method of Steelman and Pohley [10] provided this tool (still used in the late 1990s for the batch analysis of FSH products for therapy) and this aided purification and structural analysis including the recognition of the sub-unit structure of FSH though the precise elucidation of the oligosaccharides and the significance of glycosylation heterogeneity remained for future analysis.

As a result of the initiatives and energy of D.R. Bangham, the twenty-sixth meeting of the ECBS in 1974 [11] was concerned exclusively (and uniquely for this multi-disciplinary body) with one group of biologicals, hormones from all over the physiological and clinical range. The Report is a key milestone in the history of endocrinology; it has a strongly evangelical flavour with emphasis on the importance of co-operation with industry and on research and training combined with a growing awareness of the significance of protein hormone heterogeneity and, for the first time, guidelines on the diagnostic uses of hormone immunoassay.

In endocrinology, this was very much the era of immunoassay which originated in the discipline with, once again, insulin being the prototype which the pioneers, S. Berson and R. Yalow used. This technique allowed the measurement in biological fluids, and other media, of peptide hormones with a sensitivity and precision unrivalled by biological assays. It catapulted endocrinology forward but did create, or expose, standardisation problems which the traditional and accepted precepts and practices of the ECBS were only partially able to tackle and solve.

This era saw the ECBS of WHO becoming much more closely concerned with the consideration and provision of standards for in vitro diagnostic assays in endocrinology. The main problem which was only dimly, and intermittently, recognised was that the two analytical approaches – biological, immunological – were specific for, or directed at, the two separate feature of the molecules, the functional and the structural, which though they often coincided, did not necessarily do so. This led to a number of confusions, not to say conflicts, some of which are still unresolved. International Standards were for a time categorised as being «for bioassay» or «for immunoassay» a deliberate expression of the problems that were encountered in practice when one standard was used for different analytical objectives and a largely unconscious expression of the structure/function divide.

One is reminded of Heisenberg again: «measurement is a compromise and you, the experimenter, must choose what to measure and accept the consequences».

The Last 25 Years: 1973-1998

Based on these solid foundations, endocrinology was in a position of strength when the next major phase of progress in the manufacture of biologically important proteins arrived and changed everything – the era of biotechnology arising from the application of recombinant DNA (rDNA) technology. The first rDNA product to be authorised for use in human subjects was, once again, insulin – human insulin from Eli Lilly (in the US but first licensed in the UK) – in 1982. The procedures for ensuring its purity, potency and safety evolved from those used for the previously licensed naturally-derived materials. The additional concern of the potential hazard of any residual DNA of bacterial (and later of yeast and mammalian) origin were of course addressed and the ECBS drew up guidelines for industry and regulators.

At the same time techniques for purification of the products («down-stream processing») and the application of chemical methods of analysis during processing and to the testing of the end product both at the bulk and finished product stages underlined the relentless march of chemistry and the diminished role of biological assay and standardisation.

This process was well illustrated by experience with Growth Hormone. In 1985, rDNA-derived growth hormone (GH, somatropin) first became available, fortuitously coinciding with the initial revelation of the appalling consequences of the contamination of some batches of pituitary-derived GH with the prion agent that produced, after a long incubation period, the human form of the spongiform encephalopathy group of neurological diseases, Creutzfeld-Jakob Disease (CJD). Then, as now, there was no test for this contaminant in biological medicines.

The rDNA preparations of GH from a number of manufacturers soon filled the clinical need and indeed led to a widening of the clinical indications as material became relatively plentiful. At first, the new rDNA products were regulated and controlled and standardised like the previous pituitary materials but it soon became apparent that this new generation of GH products was of a purity far higher than previously with a low level of impurities (particularly the product-related peptides) that could be identified and quantified with a much higher degree of precision and reliability.

By 1990 the question had arisen: had the methods of manufacture and chemical analysis reached a stage of development and reliability that biosynthetic GH, a protein with 191 amino acids and molecular weight over 20kDa, could now cease to be regarded as a «biological»? An international collaborative study was therefore designed involving the manufacturers, control laboratories and a number of academic centres to test the ability of a range of functionally-directed (in vivo and in vitro bioassays) and structurally-directed (reversed-phase and size-exclusion HPLC, immunoassays, receptor binding assays) approaches to confirm the identity and to measure the potency and quantify the impurities in rDNA GH. The results of this important study, designed and co-ordinated by scientists at the WHO International Laboratory at the National Institute for Biological Standards and Control in the UK, were clear [12]; not only were the physicochemical methods (e.g. SE-HPLC and RP-HPLC) equivalent to the in vivo bioassay which had been the mainstay of the quality control and standardisation of this hormone, they were superior to it in precision, specificity and reliability. Other in vitro alternatives, whether cell-based bioassays or binding assays based on receptors or antibodies, added little and in many instances gave misleading and unreliable information.

It was particularly important to test in vitro bioassays since these have been long advocated as the methods destined to replace the in vivo approaches to the control and standardisation of protein hormones and considerable effort has gone into their development. For GH they failed to meet appropriate criteria of specificity and reliability and underline the conclusion I have referred to elsewhere [13, 14] that in moving from a biological to a chemical the in vitro bioassay is unnecessary and inadequate as an intermediate stepping stone; the move from in vivo assay to a chemical approach should be made without a transition except those dictated by necessary caution and the need to ensure public health. In this regard the continued use of the so-called *«bioidentity test»* is an unwarranted and, as I have also emphasised previously [13, 14], an unscientific and even unethical avoidance of the real issues underlying the evaluation of structure and function of protein hormones.

Returning to the criteria of potency, purity and identity (authenticity) which collectively determine the safety and efficacy of a protein hormone in clinical use, what are the alternatives to animal based bioassays? This is a question that has been recently restated and examined under the «3 Rs» title: *Replacement, Reduction and Refinement* of animal-based testing for biologicals [15, for example]. Once more, insulin provides a telling illustration of the problem and its solution: the 4th IS of insulin was established with unparalleled precision (95% confidence limits of ± 1%) but required the use of 23,000 mice and 4,5000 rabbits. It served well for almost 30 years (1958 to 1987) but being a 50:50 mix of bovine: porcine insulin it could not serve as a standard for the new generation of «monocomponent» human insulin products controlled by physicochemical assays. It was replaced by the 5th IS now (and the first time) of *human* insulin whilst maintaining the International Unit first set with an impure mixture of animal insulins in the 1920s. Although it is now possible to express the potency and ampoule content of the modern generation of insulin products in mass (or molar) units the IU is still used to obviate the danger to patients that might follow a change-over.

For glycoprotein hormones such as FSH and erythropoietin where the glycosylation pattern determines clearance in vivo, bioavailability and pharmacokinetics there is no place for in vitro bioassay methods. One of the challenges currently is to devise chemical approaches to reflect the oligosaccharide structures in a manner that gives physiologically and clinically relevant information; bioassays in vitro cannot do this.

I mentioned above that the immunoassay era brought challenges in the diagnostic area in the decades from the 1970s onwards that the ECBS tackled, only partially successfully. The main deficiencies revealed by the ECBS approach to diagnostics were: there was (and is) no regulatory framework for these except for a narrow range of tests (e.g. those done on blood donations to ensure microbiological safety) [16]. There are no (or few) experts on diagnostics (e.g. trained clinical chemists) who attend the ECBS and a feeling in some quarters that this area was outside its remit. In effect, the Committee may have been too remote from the issues experienced in diagnostic pathology laboratories. In endocrinology this was revealed by the decision on the IS for pituitary FSH established in 1986 against the advice of experts [17].

These problems and dilemmas can be illustrated by an examination of FSH once more. In the late 1960s it became established that the glycoprotein hormones consisted of two sub-units, alpha and beta, which separately are inactive in biological assays but have structures that can be detected in immunoassays. International Standards for these subunits were prepared and International Units were allocated to them. These caused confusion as they could not be related to the IUs of the parent hormone and the protein subunits had no intrinsic biological activity in their own right. Other attempts to resolve, or evade, the issues e.g. the use of International Reference Preparations (IRP) or International Reference Reagents (IRR) provided no better than temporary respite and may have added to the confusion (and gave ammunition to certain disbelievers and malcontents who attempted to undermine the work of the programme). These issues were debated in detail in the ECBS in the late 1980s [18] without ever being satisfactorily resolved.

The Next 25 Years: 1998-2023

If, perchance, I am asked back in 2023 to give a talk on the role of WHO in biological standardisation in endocrinology, I will be able to say that for hormones – and also for other regulatory peptides such as cytokines and growth factors – biological standardisation is a «done deal» – a thing of the past. All such proteins used in medicine will be sufficiently well-defined chemically to ensure their safe and efficacious use in the clinic. At the moment (1998) control and standardisation of glycoprotein hormones such as FSH and erythropoietin are still in the position of needing an in vivo biological assay since chemical techniques are still being evaluated for their ability to assess the molecules *adequately* and the connection between the functional and structural elements is still not completely clear [19]. But I am sure that in 2023 it will be.

References

1 Strachey, Lytton: Eminent Victorians (first published 1918); Penguin Books 1948. (preface)

2 Carr EH: What is history? (first published 1961); 2nd edition. Penguin Books 1990.

3 Jeffcoate SL, Schild GC: International Biological Standardization. WHO Drug Information 1991;5:41-42.

4 League of Nations Health Organisation: The Biological Standardisation of Insulin. (CH 398) Geneva, 1926.

5 League of Nations Health Organisation: The new International Insulin Standard and the redefinition of the existing unit in terms thereof. Quarterly Bulletin, Geneva, 1935;V:583-658.

6 Jeffcoate TNA: The Aschheim-Zondek Reaction. MD Thesis, University of Liverpool, 1931.

7 World Health Assembly, Geneva, July 1948. Section 12.1.9. International Standards.

8 Outschoorn AS: The Biological Standardization programme of the World Health Organization. J Biol Stand 1973;1:203-213.

9 Finney DJ: Statistical Method in Biological Assay. 1st edition, London 1952. Charles Griffin.

10 Steelman SL, Pohley FM: Assay of the follicle stimulating hormone based on the augmentation with human chorionic goandotropin. Endocrinology 1953;53:604-616.

11 World Health Organization Expert Committee on Biological Standardization (WHO ECBS). Report of the 26th Meeting. WHO, Geneva, Technical Report Series No 565, 1975.

12 Bristow AF, Jeffcoate SL: Analysis of therapeutic growth hormone preparations: Report of an inter-laboratory collaborative study on growth hormone assay methodologies. Biologicals 1992; 20:221-231.

13 Jeffcoate SL : The role of bioassays in the assessment of recombinant proteins, in Brown F, Lubiniecki AS (eds): Genetic Stability and recombinant Product Consistency. Dev Biol Stand. Basel, Karger, 1994, vol 83, pp 159-164.

14 Jeffcoate SL: 1966 The role of bioassays in the development, licensing and batch control of biotherapeutics. Trends in Biotechnology, 1996;14:121-124.

15 Brown F, Cussler K, Hendriksen C (eds): Replacement, Reduction and Refinement of Animal Experiments in the Development and Control of Biological Products. Dev Biol Stand. Basel, Karger, 1996, vol 86.

16 Jeffcoate SL, Lort MC: Unpublished Report to the National Biological Standards Board, South Mimms, UK, 1989.

17 World Health Organization Expert Committee on Biological Standardization (WHO ECBS). Report of the 37th Meeting. Technical Report Series, Geneva, 1987;760:24-25.

18 World Health Organization Expert Committee on Biological Standardization (WHO ECBS). Report of the 37th, 38th, 39th and 40th Meetings. Technical Report Series, Geneva, 1987, n° 760; 1988, n° 771; 1989, n° 786; 1990, n° 800.

19 Jeffcoate SL: Standardization of human follicle-stimulating hormone for therapy and diagnosis: Report of an International Workshop, NIBSC, UK, 21-22 March 1996. Clinical Endocrinology, 1997;46:527-529.

Dr. S.L. Jeffcoate, P.O. Box 188, Dorking, Surrey RH4 2FE, U.K.

Brown F, Griffiths E, Horaud F, Schild GC (eds): A Celebration of 50 Years of Progress in Biological Standardization and Control at WHO.
Dev Biol Stand. Basel, Karger, 1999, vol 100, pp 49-55

Biological Standards in Haemostasis and Thrombosis

D.P. Thomas

Division of Haematology, National Institute of Biological Standards and Control, Potters Bar, Herts., U.K.

Key words: Standards, haemostasis, thrombosis.

The provision of standards for products derived from human blood is a relatively recent development in the history of biological standardization, and it was only when purified plasma proteins were produced for the treatment of inherited coagulation defects, such as haemophilia, that a need arose for standards against which the potency of therapeutic concentrates could be assessed. As in most fields of biology, there has been an enormous growth of knowledge in the past twenty years, and there are today biological standards for many of the purified proteins used in the treatment of such haemostatic defects as haemophilia, as well as standards for fibrinolytic and anticoagulant drugs used to control and treat the inappropriate coagulation called thrombosis. Indeed, attempts to standardize heparin, perhaps the most important anti-coagulant drug, antedate even the formation of the Expert Commitee on Biological Standardization [1].

The marked growth in understanding of the mechanism of blood coagulation in the 1960s was accompanied by a dramatic improvement in the care of patients with inherited clotting defects. The commonest of these defects is classical haemophilia, a disease recognized since ancient times, and which today is understood to be due an inherited deficiency of the clotting factor known as Factor VIII. The development by the plasma fractionation industry some thirty years ago of purified concentrates in a stable freeze-dried form meant that effective treatment for haemophilia finally became a reality. As a direct consequence of this important therapeutic development, an internationally agreed unit was needed for measuring Factor VIII in purified concentrates. For many years this unit was defined as that amount of Factor VIII present in one ml. of fresh blood. However, this value varied accordingly to age, sex and even geographical location of the blood donor. Clearly, to define a unit of Factor VIII solely in terms of fresh normal plasma is an inadequate basis for standardization on a world-wide basis, when so-called «normal» values can vary by 20% or more, depending on the donors. To overcome this difficulty, a standard for the Factor VIII molecule as measured in therapeutic concentrates was established in 1970, primarily in response to the growing international trade in Factor VIII concentrates [2]. Subsequently, it became apparent that Factor VIII in these concentrates was not quite the same as the molecule present in normal plasma. A standard for Factor VIII in plasma was therefore established, following the time-honoured practice in biological standardization of comparing «like with like». In some countries, a national working standard for Factor VIII in plasma is provided to specialized haemophilia centres, where it is used for laboratory control in the day-to-day treatment of patients. In the UK, the scheme has been highly successful, and has been running for some 25 years.

There are now several other purified clotting factors beside Factor VIII for which an international standard has been prepared (Table 1). For example, the clot-

Table 1: Coagulation standards.

Material	Code n°	Units/ampoule	Description
Antithrombin concentrate	96/520	Functional 4.7 IU Antigen 5.1 IU	2nd.IS
Antithrombin, plasma	93/768	0.85 IU	2nd.IS
Factor II, VII, IX, X plasma II VII IX X	94/746	 0.93 IU 1.25 IU 0.90 IU 0.95 IU	2nd. IS
Factor II, X concentrate II X	84/683	 9.4 IU 11.1 IU	2nd.IS
Factor VII concentrate	97/592	6.3 IU	1st. IS
Factor VIIa concentrate	89/688	5130 IU	1st. IS
Factor VIII:C concentrate	97/616	8.5 IU	6th. IS
Factor VIII-related activities in plasma VIII: C VIII: Ag vWF: RCoF vWF: Collagen binding vWF: Ag	97/586	 0.57 IU 0.89 IU 0.73 IU 0.79 IU 0.78 IU	4th. IS
Factor IX concentrate	96/854	10.7 IU	3rd. IS
Factor IXa human	97/562	11.0 IU	1st. IS
Fibrinogen, plasma	89/644	2.4 mg	1st. IS
Protein C, plasma	86/622	0.82 IU	1st. IS
Protein S, plasma	93/590	Total 0.90 IU Free 0.90 IU Function 0.90 IU	1st. IS
Thrombin, α-type, human	89/588	100 IU	2nd. IS
Thrombin, human	70/157	100 IU	1st. IS

ting factor whose absence is responsible for the other main variety of haemophilia, namely Factor IX deficiency, representing some 15% of all haemophiliacs, has been purified and a separate international standard established. In the past, a complex of Factors II, IX & X – the so-called «prothrombin complex» – was used for the standardization of Factor IX, but with the advent of monoclonal and recombinant technology, highly purifed Factor IX concentrates have been developed which require their own standard. The general agreement about Factor VIII and Factor IX standards, and how they are defined, is due to close scientific liaison between scientists in many countries.

The United States is the major exporter of Factor VIII and Factor IX concentrates for the treatment of both types of haemophilia, and it was therefore particularly important that American manufacturers agreed to adopt international units when selling their products. This was achieved after many discussions over several years with manufacturers and the CBER division of the FDA, and WHO Units for these concentrates are now the acknowledged international «gold standards». Factor VII is the most recent addition to the list of purifed clotting factors for which an international standard is available, and there are several other clotting factors and inhibitors for which standards are available, in both plasma and concentrate form (Table 1).

A more difficult issue is the provision of WHO working standards for everyday use by manufacturers. Traditionally, the international standard has been available in only limited quantities, and over the years the practice has been for manufacturers to calibrate their own house standard against the international standard. However, with new techniques, it may now be possible to provide a common working standard for all manufacturers of Factor VIII concentrates, as has already been achieved in some countries on a national basis. It is an attractive prospect for manufacturers world-wide to use a common working standard, obviating the need for secondary calibration of their own working standard from the primary standard. While the system of secondary calibration has worked well enough over the years, in my view it would be preferable to have a common working standard for all, thus eliminating an important source of assay variation [3]. For example, the technology exists to fill Factor VIII concentrate into, say, 150,000 rubber-capped vials for standardization purposes. It remains to be established whether such a working standard has sufficient stability to serve as a WHO International Working Standard for Factor VIII concentrate for several years. The encouraging experience in the United States with the MEGA 1 standard, used successfully as a working standard by all US manufacturers for over 10 years, suggests that this is indeed the case. There will of course still be a need for the primary international standard, stored in glass-sealed ampoules under optimal conditions.

There are other important areas in the field of haemostasis and thrombosis that rely on biological standards for the production of therapeutic products. In the fibrinolysis field, there are standards for streptokinase, plasmin, urokinase and tissue plasminogen activator, among others (Table 2). Such standards play a vital role is ensuring that drugs are manufactured to a consistent potency, so that clinicians can rely on the relative predictability of a therapeutic dose. For example, an international standard for the important clotting enzyme thrombin has existed for many years, and while this enzyme is rarely used therapeutically, it is nevertheless an essential component of many laboratory assays and kits, and a recognized international unit has been of great help in standardizing the potency of thrombin.

Table 2: Fibrinolysis and defibrination standards.

Material	Code n°	Units/ampoule	Description
Ancrod	74/581	55 I U	1st.IRP
Plasmin	97/536	5.3 IU	3rd. IS
Plasminogen activator inhibitor-1, plasma	92/6545	7.5 IU t-PA 7.0 IU u-PA	1st. IS
Streptokinase	88/826	700 IU	2nd. IS
Tissue plasminogen activator (t-PA)	86/670	850 IU	2nd. IS
Urokinase, high molecular weight (HMW)	87/594	4300 IU	1st. IS

The thrombolytic drugs streptokinase and tissue plasminogen activator are essential drugs in modern cardiology, and their successful clinical use owes much to the well-characterized international standards that have been available for assigning potency values.

The provision of standards for thromboplastins, which are needed for the control of oral anticoagulant therapy, is another area where WHO has made a highly important contribution. The role of thromboplastins in the control of anticoagulant therapy is crucial, and international standards were established in the mid-1970s. The original thromboplastins derived from human and rabbit brains have now been replaced by recombinant material, and the current international standard for recombinant human thromboplastin is held at the WHO Laboratory of the Dutch Red Cross (CLB) [4]. Equally important to the provision of thromboplastin standards is the system of calibration that was originally devised by Professor T.B.L. Kirkwood, when he was at NIBSC. He introduced a system whereby thromboplastin activity is expressed in terms of the International Sensitivity Index (ISI), relating a particular thromboplastin to the International Standard [5]. Using this calibrated preparation it becomes possible to evaluate any treated patient's plasma in terms of a single numerical system, the so-called International Normalized Ratio (INR). This system of calibration has now been generally accepted worldwide, and is recognized as an essential part of oral anticoagulant therapy with coumarin drugs.

The first, and still one of the most important, biologicals in this field for which an international standard exists is the anticoagulant heparin. A provisional standard was prepared as long ago as 1942 [1], and confirmed in 1947 at the first meeting of the Expert Committee on Biological Standardization. Heparin, or unfractionated heparin as it is called today, is therefore one of the longest serving biological standards. Over the past 50 years, the standard has been replaced several times, and is still one of the most widely used (Table 3). However, the WHO heparin standard exemplifies the need for an agreed unit, for unfortunately no such agreement exists for unfractionated heparin worldwide. Many years ago, the United States Pharmacopoeia established its own unit for heparin, and this has now diverged from the international unit by some 10%. With a view to resolving this unsatisfactory situation, the Biologics Unit at WHO recently convened a meeting to review heparin standardization, and it is hoped that a measure of harmonization

Table 3: Anticoagulants and other reference preparations.

Material	Code n°	Units/ampoule	Description
Heparin, porcine mucosal	97/578	2031 IU	5th IS
Heparin, low molecular weight Anti-Xa Thrombin inhibitor	85/600	 1680 IU 665 IU	1st. IS
Prekallikrein activator (PKA), human	82/530	85 IU	1st. IS
Platelet Factor 4 (PF4), human	83/505	400 IU	1st. IS
β-thromboglobulin (BTG), human	83/501	500 IU	1st. IS

between the two units can eventually be achieved. However, the main difficulty is the different approach to standardization taken by WHO and the pharmacopoeias. WHO has traditionally taken the view that the potency of a standard is best expressed by taking the mean of all valid assays used in a collaborative study. In contrast, the pharmacopoeias have always used a defined assay, and assign potency values solely on the basis of the pharmacopoeial assay. A similar problem exists with Factor VIII, where the European Pharmacopoeia now has a defined assay (chromogenic substrate), and assigns a unit for Factor VIII solely on the basis of this assay. In other parts of the world, Factor VIII is usually assayed by the one-stage clotting assay, which can give a different potency figure to a sample of Factor VIII. This differing approach to biological standardization between WHO and the pharmacopoeias is clearly unsatisfactory and is in urgent need of resolution.

Low molecular weight heparin is an interesting example of how an improved drug can develop from an existing one. Over the past 10 years or so, low molecular weight heparin has become increasingly preferred to standard heparin for many clinical indications, and now promises to replace the older drug. However, at an early stage in development of the newer heparin it became apparent that the existing heparin standard was inappropriate for standardization, largely because of lack of parallelism in biological assays, and so a separate low molecular weight standard was established in 1986 [6]. This standard has its own international units, and unlike the unit for unfractionated heparin, is accepted worldwide. However, this is not the end of the story, because as more low molecular weight heparins are introduced into the market, the issue that now has to be resolved is whether a single generic standard is adequate for all low molecular weight heparins, which vary slightly in their method of manufacture, or whether, as some would claim, a whole series of different low molecular weight heparin standards will eventually be needed.

The story of low molecular weight heparin also illustrates the limitations of the defined pharmacopoeial assay. If heparin had only been measured by the pharmacopoeial assay, the peculiar characteristics of low molecular weight heparin would never have been discovered [6]. In my view, it is essential when defining the characteristics of any biological that an inclusive approach is adopted, and the role of new assays allowed for, perhaps reflecting different, but equally important, biological activities. The WHO approach to standardization has always allowed for such developments, in that assigned potencies represent the mean of all valid assays, and are not restricted to one defined assay. The example given for heparin and low molecular weight heparin, and the need for a broad approach to standardization, could be multiplied many times. The important point to emphasise is that biological standardization must take place within the context of the scientific field for which standards are being provided, and it follows that those who have this responsibilty need to retain their scientific credibility by remaining actively involved in related research. This crucial point was recognized early on, and in the UK the provision of standards for the international community was for many years the responsibility of the Medical Research Council.

The timely provision of standards in a fast developing field is a considerable challenge, and the need for a standard must be identified at an appropriate stage in the development of any new biological. With the vital help of statistical colleagues, a collaborative study has to be organized and candidate materials examined. It is appropriate here to acknowledge the invaluable help given by the world-wide scientific community, who participate in the often extensive studies required to establish a standard. Without their ready and freely-given cooperation, no standard could ever be established. There are also the manufacturers who usually provide the candidate materials for a standard, and who must accept that if their particular product is chosen for an international standard, this is not a reason for promotional advertising. In my experience, most manufacturers are willing partners in this endeavour, for they recognize fully the importance of an agreed international standard, especially in an increasingly global economy.

Finally, it should be emphasised that WHO standards can only operate on the basis of consensus, and the acceptance of an international standard depends to a great extent on the quality of the scientific input into the collaborative study, and acceptance by international professional societies that WHO standards have an essential role to play in the advancement of their subject. The International Society of Thrombosis and Haemostasis (ISTH) has been of immense help in promoting the widespread use of international standards, and there has been a long and fruitful collaboration between ISTH and the WHO Laboratories and Collaborating Centres for Biological Standardization. The Society, now recognized by WHO as an NGO, has appointed a liaison officer for standardization issues, acting between the Biologicals Unit at WHO in Geneva, and the Headquarters of ISTH in the United States.

What of the future? As the recent scientific review commissioned by NIBSC concluded, progression to better characterized biologicals will help eliminate the inherent variability that has been a central problem in the control of biologicals [7]. At present, new developments raise more questions than can be answered about predictablility and about how, and how reliably, new products can achieve their effects, especially in the burgeoning field of biotechnology. The replacement of some products derived exclusively from human blood by recombinant equivalents, such as recombinant Factor VIII, means that new standards and assay methods have had to be developed. While recombinant blood products are very similar to plasma-derived products, it is also becoming clear that they are not identical, and this lack of identity can cause assay problems that have to be resolved. Continuing problems of variability in the potency of Factors VIII and IX concentrates underline the importance of improving our understanding of the antigenic properties of these molecules, and the need for improved test methodology. Similarly there is active development of recombinant variants of thrombolytic drugs, with variants that have already been developed to prolong the in vivo half-life and drug dissolving effect of the agents. This work has to be carried out in close collaboration with the manufacturers, who are in fact essential partners in the process of establishing new and appropriate standards for their products. While many products developed by recombinant technology will have a unique composition and each one will have a slightly different spectrum of biological effects, a standardization package specific to the purity and consistency of their own unique product will still need to be worked out with the manufacturer. Where there are families of products that have similar clinical activity, generic standards will be needed even if these do not fully characterize a specific product, and each product might require its own array of physico-chemical tests. In the haemostasis and thrombosis field it will be easier to establish such generic standards, where bioassays can usually directly measure the desired biological effect, namely changes in a clotting or fibrinolytic assay.

Finally, great strides have been made in the past 50 years in the provision of international standards for the control of biological substances, and with the new challenges that are continually developing, there seems little doubt that the next 50 years will be equally exciting.

Summary

International standards are used to assess the potency of a variety of biological drugs employed in the field of haemostasis and thrombosis. These include therapeutic concentrates for the treatment of inherited coagulation defects, such as haemophilia, thrombolytic drugs such as streptokinase for the treatment of heart attacks, and anticoagulant drugs such as heparin. Some of these standards have existed for over 25 years, and they continue to play a vital role in ensuring consistent potency, and hence safety, of manufactured drugs. The growing complexity of new products, particularly in the area of recombinant technology, represents a continuing challenge for the field of biological standardization.

References

1 Bulletin of the Health Organization, League of Nations, 1943/44;10 (2):151.

2 Bangham DR, Biggs R, Brozovic M, Denson KWE, Skegg JL: A biological standard for measurement of blood coagulation Factor VIII activity. Bulletin of the World Health Organization, 1971;45:337-351.

3 Barrowcliffe TW: Standardization and Assay. Seminars in Thrombosis and Haemostasis, 1987; 19:73-79.

4 WHO Expert Committee on Biological Standardization: 47th Report. WHO Technical Report Series, 1998;N° 878:13.

5 Requirements of Thromboplastins and Plasma used to Control Oral Anticoagulant Therapy. WHO Technical Report Series, 1983;N° 687:Annex 3.

6 Thomas DP: Biologicals, Standards and Heparin. Thrombosis and Haemostasis, 1989;62:648-650.

7 Biological Standardization and Control: A scientific review commissioned by the UK National Biological Standard Board (NBSB). World Health Organization, Geneva;1997:(WHO/BLG/97.1).

Dr. D.P. Thomas, The Old Barn, North Green, Kirtlington, Oxford OX5 3JZ, U.K.

Brown F, Griffiths E, Horaud F, Schild GC (eds): A Celebration of 50 Years of Progress in Biological Standardization and Control at WHO.
Dev Biol Stand. Basel, Karger, 1999, vol 100, pp 57-63

Cell Substrates: Lessons Learned and Challenges Remaining

J.C. Petricciani

Palm Springs CA, USA

Key words: Cells, contamination, viral contaminants, biological products, adventitious agents.

Abstract: The history of cell substrates for the manufacture of biological products is directly related to a series of technical advances and challenges to the status quo, some of which were accepted quickly while others took more than a decade to resolve. The development of cell culture techniques in the 1950s opened the door to the manufacture of a wide range of biological products. The first major challenge occurred in the late 1960s when human diploid cells (HDCs) were developed and proposed as an alternative to primary cell cultures for the production of live viral vaccines such as polio, which up to that point had been produced in primary cells of various species. In the 1970s, attention was focussed on the use of continuous cell lines (CCLs) for the production of non-replicating biological products such as interferon (IFN). The next significant technical advance and challenge was the development of recombinant DNA and monoclonal antibody technologies in the 1980s, both of which required the use of CCLs. Although most of the issues relating to CCLs in the manufacture of biological products have been resolved, issues related to their use as substrates for live viral vaccines remain to be fully addressed.
Those experiences in the past teach us clearly that a system in which regulatory authorities, industry, and the general biomedical community cooperate in finding solutions to problems and in reaching consensus on issues raised by technical advances is ultimately in everyone's best interest. The World Health Organization has played a major role in that regard, and it should continue to provide leadership in this area.

INTRODUCTION

As the title suggests, this paper presents an overview historical perspective on cell substrates used in the manufacture of biological products, and it examines two issues remaining to be resolved. Only the major points of the history of cell substrates over the past four decades are highlighted, since the details have been discussed elsewhere [1].

The first point to note is that cell substrates have been a recurring focus of attention during the past 50 years, often with a great deal of anxiety attached. The two core issues that continue to be the focus of attention even today are the potential for transmitting infectious diseases and/or cancer.

The first major decision regarding cell substrates was, in fact, to avoid the theoretical risk of transmitting cancer through live viral vaccines by restricting the acceptable types of production cells to those from primary tissues, and prohibiting the use of cells derived from human tumours, such as HeLa, for such purposes [2].

The first viral contaminant identified in a biological product which had potential major public health significance was SV40, an endogenous virus of Rhesus monkeys [3]. The public health issues related to the use of primary Rhesus monkey kidney cell cultures to produce polio vaccines and the discovery that SV40 was tumourigenic in hamsters [4]. There was therefore justifiable concern that the SV40 contaminant in polio vaccines might lead to cancer in the vaccinated children. That concern was heightened with the finding that SV40 could transform normal human diploid fibroblasts, WI-38, in vitro.

There were two major responses to the SV40 incident. First, the National Cancer Institute initiated a long-term follow-up study of a cohort of those children who had been exposed to SV40. After 20 years, there was no evidence of an increased frequency of cancer [5]. A more recent 30-year follow-up study confirmed the conclusions of previous reports that there is not a significant correlation between SV40 exposure during the neonatal period and certain cancers [6]. Second, the cell substrate was switched from Rhesus to African Green monkey kidney since SV40 is not endogenous for that species.

Nevertheless, concern about endogenous viruses in nonhuman primates continued as more and more latent viruses were identified. The potential for serious consequences was accentuated in the 1970s with the deaths of a number of persons who came into direct or indirect contact with African monkeys that were infected with what became known as the Marburg virus [7].

Two solutions to the problems associated with using primary cells as substrates for the production of biologicals were advanced. One was to develop closed and carefully monitored colonies of animals, and in some cases specific pathogen-free colonies, from which primary cell cultures could be derived. The other approach was put forward by Dr. Hayflick and his colleagues in the 1960s [8]. They developed an alternative to primary monkey kidney cells, namely, the WI-38 human fibroblast cell line which has normal phenotypic characteristics, such as a finite life, a diploid chromosomal constitution, and a lack of tumourigenic potential. In addition, a variety of tests showed no evidence of viral contamination. However, WI-38 was slow to be accepted in the U.S., primarily because of Dr. Albert Sabin's position that they might contain latent human viruses pathogenic for humans, especially a human leukaemia virus for which there were no tests [9]. Obviously, if one doesn't know how to test for an agent that might be present, about all one can do is provide as much indirect safety evidence as possible. Eventually, that approach worked, and products derived from WI-38 were approved by FDA, but long after they had been approved in Europe. In retrospect, Dr. Sabin was correct to have been concerned about human leukaemia agents since they were discovered years later. However, he had focussed on the wrong cell type.

The next major event in the evolution of cell substrate issues came in the mid-1970s when the Namalwa human lymphoblastoid cell line was proposed by the Wellcome Research Laboratories as a source of interferon (IFN) for human clinical studies [10]. IFN from malignant human lymphoid cells presented serious questions of safety and provided the first direct challenge to the rather firmly established concept that acceptable cell lines had to possess HDC characteristics of a finite life, a normal cytogenetic profile, and not produce tumours. After considerable discussion [11], lymphoblastoid IFN was used in clinical studies and was eventually licensed in several countries. That research and development effort laid the groundwork for approaches that would be taken in the 1980s to ensure the safety of products derived from a variety of CCLs, such as the VERO cell line which was used by Merieux to manufacture more economical inactivated and live polio vaccines [12].

Subsequently, biotechnology led many groups to pursue CCLs as expression systems for new products. CHO became the workhorse for rDNA products, and hybridomas were used for monoclonal antibodies. However, there was still a great deal of concern relating to the safety of products for human use derived from various CCLs. As in the past, the central issue was the theoretical possibility that products derived from CCLs could transmit cancer and infectious agents to humans. These «new» problems were really nothing more than variations on the general themes of the past. The actors were different, and they may have added some nuances to the drama, but the basic themes were very much the same as they were in the 1950s.

Rather than focussing on cell characteristics such as tumourigenicity and chromosomal constitution, attention was finally shifted to risks that might be associated with three potential contaminants from the cell substrate: i) viruses; ii) DNA; and iii) transforming proteins [13]. A number of meetings were held to discuss those issues, and a growing consensus emerged that CCLs should, in principle, be acceptable as substrates for biologicals production [13, 14]. The manufacturing process then must be shown to eliminate potential contaminating viruses pathogenic for humans and to reduce DNA to acceptable levels and/or to eliminate its biological activity [14]. Transforming proteins were considered essentially trivial risks.

The most significant issue from a public health and safety point of view has been, and continues to be, viral and viral-like contaminants. In that regard, it is perfectly clear that primary tissues and primary cells have been the source of several problems that occurred in the past (Table 1).

As already mentioned, polio vaccines carried SV40 from primary Rhesus monkey kidney cells. Fortunately, there is no evidence to suggest that the children who received SV40 in polio vaccines were at greater risk for cancer than those who did not get SV40 [6].

Table 1: Transmissions of infectious agents from cells and tissues.

	Product	Cell / Tissue	Agent / Disease
P	Polio vaccine	1° Monkey kidney	SV40/?
	PPF	Human plasma	Hepatitis B
A	Transplants	Human cornea	Rabies
		Human dura mater	Prions/CJD
S	Growth hormone	Human pituitary	Prions/CJD
T	Factors VIII & IX	Human plasma	HIV/AIDS Hepatitis A, B, C
N	Bovine tissues/Organs	Beef (food)	BSE/?
O	Blood & derivatives	Human cells/Plasma	Prions/CJD
W	Xenotransplants	Pig organs	Retroviruses/?

Human Plasma Protein Fraction (PPF) from one producer transmitted hepatitis B in the 1970s, probably as the result of a manufacturing flaw, and some deaths occurred. Human corneas derived from a cadaver transmitted rabies after they were transplanted because the medical history of the cadaver was ignored. Similarly, growth hormone derived from the pituitaries of human cadavers led to the development of Creutzfeld-Jacob disease (CJD) in many of the children who received that product. In addition, dura mater transplanted from cadavers has transmitted CJD. Factor VIII (Antihaemophilia Factor) and Factor IX Complex derived from human plasma in the 1960s transmitted hepatitis viruses because neither tests for donor screening nor virus inactivation procedures for those products were available. When HIV entered the plasma supply in the late 1970s or early 1980s, it also was transmitted to haemophilia patients in Factors VIII and IX, with over 4,000 of them in the U.S. having developed AIDS since that time. Currently, we are worried about: (i) the agent responsible for bovine spongioform encephalopathy as a contaminant of beef, especially in Europe; (ii) prions as a potential contaminant of blood and blood products; and (iii) retroviruses as potential pathogens in transplants of organs from various species to humans.

In contrast to these examples of actual transmission of disease, or concern about possible transmission, closed colonies of animals from which primary cell cultures are derived, diploid and continuous cell lines, and raw materials can be tested extensively to rule out the presence of potential pathogens for humans. The ability to pre-test has provided a substantial improvement in the quality of cell substrates and it is the single most significant factor that has resulted in the absence of any documented case of infectious disease transmission by any product of which I am aware that was derived from such well-characterized and tested cell substrates.

Current Issues

While many cell substrate issues have been resolved or are on their way towards resolution, there are some that remain to be addressed. Two of those are highlighted here.

The sophistication and sensitivity of test systems are continually increasing. That, in turn, raises the question of how we are going to apply them in a rational way to the routine quality control of new products. Two examples of this issue are the vastly increased test sensitivity that is now available for specific genetic sequences such as reverse transcriptase and for host cell protein. Balanced against the temptation to require state-of-the-art methods just because they exist must be a recognition that there is a fundamental link between safety and purity, and that the driving force should be safety. In other words, the objective should be to achieve a degree of purity that provides every reasonable assurance of safety. Obviously, state of the art tests are often needed to provide assurance of safety. But the general principle of manufacturing products that are pure enough to be safe should be applied, rather than pursuing purity as an abstract goal in and of itself. Fortunately, a substantial body of clinical experience exists in both the human and veterinary fields from which safety information can be obtained to help address cell substrate safety issues, and this should not be neglected as new issues arise.

The second, and most significant cell substrate issue that remains to be addressed is the use of CCLs to produce live viral vaccines. During the past two decades we have learned through specific experiments and practical experience with a variety of products how to deal rationally with CCLs in the manufacture of non-replicating biologicals and inactivated vaccines. However, the use of CCLs to produce live viral vaccines has been limited to VERO cells for polio. The safety concerns again center around the possibility of transmitting cancer and/or viruses pathogenic for humans. The specific issue is whether a live viral vaccine replicating in a CCL could incorporate harmful nucleic acid sequences from the cell into the virus as an epigenetic element or directly into the viral genome itself. If either of those possibilities occurred, then the viral vaccine could act as a vector and transmit them to the vaccinees. In the latter case, for example, an oncogenic element from the CCL could theoretically be incorporated into the viral genome and result in an oncogenic virus that could transform human cells. These are important issues that require careful consideration as live vaccines manufactured in CCLs are considered. The scientific issues need to be clearly articulated and agreed upon by regulatory authorities and industry, then specific experiments to answer the scientific questions should be defined and agreed upon, and finally the results must be evaluated and discussed in public.

Major Contributors to Advances in Cell Substrates

Many individuals, such as Drs. Albert Sabin, Jonas Salk, and the members of the 1986 WHO Study Group, have made significant contributions to this area; but the individuals and organizations listed in Table 2 have been particularly instrumental in pushing the field to the next level of technology which then opened the door to the manufacture of products that were not previously possible.

Summary and Conclusions

A number of lessons have been learned in the past century that could help as the risk that products might transmit infectious viruses and viral-like agents are evaluated. In a sense, we have been both frightened and reassured. Extraneous

Table 2: Individuals and organizations that have made major contributions to the evolution of cell substrates used in the production of biologicals.

Robert Hull	Early advocate for CCLs for vaccines
Leonard Hayflick	Developer and advocate for use of human diploid cell line WI-38
Patrick Jacobs	Developer of MRC-5 human diploid cells
Frank Perkins	Advocate for use of human diploid cells and hybridomas
Toon van Wezel	Use of microcarriers and CCLs for vaccine production
Wellcome Ltd.	Use of Namalwa cells for interferon production
Merieux	Use of VERO for polio vaccine production
Genentech	Use of CHO for hepatitis B vaccine and tPA production
Ortho, Hybritech	Use of hybridomas for monoclonal antibody production

agents have, on a few occasions, been administered inadvertently to thousands of recipients of biological products. On balance, however, considering the billions of doses and exposures that have occurred during this century, it is remarkable that there have been relatively few problems. However, we have only to think back to the early 1980s when HIV first appeared, or more recently to the difficult issues that prions present, to understand that, as Dr. Fred Robbins often pointed out, we should be prepared for surprises and not be deluded into thinking that we have all the answers or that everything is under perfect control.

If we look at the history of the actual transmission of infectious diseases by biological products, there are several points that emerge. First, problems are almost inevitable when strict control of the manufacturing process (including design, validation, and routine execution) or of the starting materials is not maintained, and when up-to-date technology is not applied. Second, a safe product can be made unsafe by the addition of a contaminated excipient (diluent, stabilizer, etc.). Third, the biggest risk of the past has clearly been associated with human agents that have contaminated primary human cells and tissues. The lesson here is that if it is possible to avoid using primary human material, it is safer and wiser to do so. Obviously, that is easier said than done, especially if one has no other choice, at least for the present, such as with whole blood and many blood derivatives.

The final point that becomes obvious from past experience is that conventional wisdom does not necessarily lead us down the right path. For example, the initial anxiety and apprehension concerning the safety of human diploid cells, or more recently of CCLs, did not prove to be well-founded. More importantly, those biases and the failure to address the issues in a coherent and expeditious manner stifled progress in dealing with basic improvements in products such as rabies vaccine, the vast majority of which continues to be produced in sheep brains, in contrast to veterinary rabies vaccines, most of which were switched long ago to CCLs such as BHK-21. Maurice Hilleman recently wrote about this issue in the context of an HIV vaccine as follows: «(the) quest... will be helped by avoiding unenlightened authorities who ooze conventional wisdom and by staying clear of doom sayers who are ever in abundance. Mustering and control of resources, acquisition of facts, creation of a strategic plan, and the practice of selective choice are the elements that are essential.» His suggestions, like so many that he has made in the past, are widely applicable, and should be remembered as new issues arise.

It is my personal belief that we are at a significant juncture in the history of biologicals where the national and regional isolationism, independence, and confrontations of the past are being replaced by engagement, consensus building, and cooperation on the appropriate regulation of biologicals. In that regard, it is clearly in the interest of health in both the developed and developing countries for WHO to exercise a leadership and coordinating role, and we should look carefully at the experiences of the past to guide us in addressing the remaining cell substrate issues as we move into the future.

Acknowledgement

Dr. John Finlayson provided particularly helpful advice and examples of past viral transmissions.

References

1 Petricciani JC: Cells, science, and health, in Hayflick L, Hennessen W (eds): Continuous Cell Lines as Substrates for Biologicals. Dev Bio Stand. Basel, Karger, 1989, vol 70, pp 3-10.

2 Hilleman M: Cell line saga – an arguement in favor of production of biologicals in cancer cells, in Petricciani JC, Hopps HE, Chapple PJ (eds): Cell Substrates. New York, Plenum, 1979, pp 47-58.

3 Shaw K, Nathanson N: Human exposure to SV40: review and comment. Am J Epidemiol 1976; 103:1-12.

4 Girardi AJ, Sweet BH, Slotnick VB, et al: Development of tumours in hamsters inoculated in the neonatal period with vacuolating virus, SV40. Proc Soc Exp Biol Med 1962;109:649-660.

5 Mortimer EA Jr, Lepow MI, Gold E, Robbins FC, Burton GJ, Fraumeni JF Jr: Long-term follow up of persons inadvertently inoculated with SV40 as neonates. N Eng J Med 1981;305:1517-1518.

6 Strickler HD, Rosenberg PS, Devesa SS, Hertel J, Fraumeni JF Jr, Goedert JJ: Contamination of poliovirus vaccines with simian virus 40 (1955-1963) and subsequent cancer rates. J Am Med Ass 1998;279:292-295.

7 Gear JSS, Cassel GA, Gear AJ, et al: Outbreak of Marburg virus disease in Johannesburg. British Med J 1975;4:489-493.

8 Hayflick L, Moorhead PS: The serial cultivation human diploid cell strains. Exp Cell Res 1961; 25:585-621.

9 Sabin A: Discussion, Session V of the International Conference on Rubella Immunization. Amer J Dis Child 1969;118:378-381.

10 Beale AJ: Choice of cell substrate for biological products. Advances in Experimental Biology and Medicine 1979;118:83-97.

11 Petricciani JC, Hopps HE, Chapple PJ (eds): Cell Substrates. New York, Plenum, 1979.

12 Montagnon BJ, Fanget B, Nicolas AJ: The large-scale cultivation of Vero cells in micro-carrier culture for virus vaccine production. Preliminary results for killed polio-virus vaccine, in Hennessen W, van Wezel AL (eds): Reassessment of Inactivated Poliomyelitis Vaccine. Dev Biol Stand. Basel, Karger, 1981, vol 47, pp 55-64.

13 Hopps HE, Petricciani JC (eds): Abnormal Cells, New Products, and Risks. Tissue Culture Association, Gaithersburg, 1985.

14 Report of the meeting of the study group on biologicals. WHO Technical Report Series 1987; N° 747.

Dr. J. Petricciani, P.O. Box 1925, Palm Springs, CA 92263, USA.

Brown F, Griffiths E, Horaud F, Schild GC (eds): A Celebration of 50 Years of Progress in Biological Standardization and Control at WHO.
Dev Biol Stand. Basel, Karger, 1999, vol 100, p 65

SESSION II

New Laboratory Tools for Standardization and Control

Co-chairs: *Alan Shaw (USA)*
and Takao Hayakawa (Japan)

Brown F, Griffiths E, Horaud F, Schild GC (eds): A Celebration of 50 Years of Progress in Biological Standardization and Control at WHO.
Dev Biol Stand. Basel, Karger, 1999, vol 100, pp 67-74

Molecular Consistency Monitoring of Oral Poliovirus Vaccine and other Live Viral Vaccines

K.M. Chumakov

Center for Biologics Evaluation and Research, Food and Drug Administration, Rockville, MD, USA

Key words: Mutant quantitation, vaccine quality control, neurovirulence test, oral poliovirus vaccine.

Oral Poliovirus Vaccine (OPV), especially its type 3 component, is prone to partial reversion to neurovirulence, requiring testing of every batch of monovalent vaccine in monkeys. We have developed a new approach to quality control of OPV production. By using mutant analysis by PCR and restriction enzyme cleavage, MAPREC, we have shown that all batches of type 3 OPV contain small amounts of U to C reversion at nucleotide 472. Batches that failed the monkey neurovirulence test (MNVT) had a higher content of mutants and were easily discriminated from acceptable batches. After conducting an International Collaborative Study, the WHO recommended MAPREC as a pre-screening test for evaluating the acceptability of type 3 OPV. Type 1 and type 2 OPV batches also contain reversions in the same RNA domain and the level of revertants depended on manufacturing conditions. We were not able to confirm previous claims by others that these mutations in type 1 and type 2 OPV increase neurovirulence. Genetically engineered constructs and high-passage samples of type 1 and type 2 Sabin strains with substantial amounts of the mutations did not possess increased neurovirulence. Considering that no type 1 or type 2 OPV batch has failed the MNVT in almost 40-year long history of OPV, we conclude that MAPREC is a far more sensitive and reliable method for monitoring consistency of OPV production than the MNVT. We propose that for type 3 OPV the MNVT might be replaced by a combination of MAPREC and a neurovirulence test in transgenic mice susceptible to poliovirus. For types 1 and 2 the MNVT might be replaced by MAPREC alone, since the animal models have not provided useful information. New data on the development of similar molecular tests for the evaluation of Mumps and Yellow Fever vaccines, as well as a general conceptual discussion of molecular consistency monitoring of live viral vaccines, will be presented.

Introduction

The stability of attenuation phenotypes of live viral vaccines is a critical factor for their safety. From the beginning of OPV development, it was recognized that OPV is prone to partial loss of attenuation markers [1]. During its manufacture and replication in the intestinal tracts of vaccinees, Sabin strains (primarily Sabin 3 strain) regained some neurovirulence for monkeys, prompting the WHO and national control authorities to require testing of every batch of vaccine in monkeys as a part of the routine lot release process [2]. The monkey neurovirulence test (MNVT) for each of the three monovalent batches that constitute one trivalent pool of OPV, requires inoculating up to 100 monkeys. The animals are subsequently sacrificed and their spinal cords and brains are subjected to histological examination. In addition to the use of a large number of valuable primates, the test possesses an inherent variability which sometimes makes interpretation difficult. For this reason the WHO set a goal of finding an alternative to the MNVT. Studies performed in the 1980s in a number of laboratories elucidated the molecular basis of neurovirulence and attenuation of poliovirus [3]. These studies showed that a mutation in the 5'-UTR of the Sabin 3 strain, that occurs regularly during virus replication in intestinal tracts of vaccinees, results in a marked increase of the monkey neurovirulence. Other mutations that distinguish a wild-type progenitor from its attenuated vaccine derivative were also identified. We wondered, therefore, whether these mutations revert and accumulate during vaccine production causing some batches to fail the MNVT occasionally. To test this hypothesis we developed an extremely sensitive method for mutant quantification: namely, mutant analysis by PCR and restriction enzyme cleavage (MAPREC), and used it to show that, indeed, there was a very close correlation between the contents of these revertants in batches of type 3 OPV and results of the MNVT [4, 5]. This report summarizes more than eight years of work with OPV and introduces a new concept for molecular consistency monitoring applicable to a wide range of live viral vaccines.

Methods

MAPREC (mutant analysis by PCR and restriction enzyme cleavage)

Aliquots of vaccine samples were extracted with phenol and SDS, and the resulting RNA was reverse transcribed into cDNA using a random hexamer primer. Single-stranded DNA produced by asymmetric PCR amplification was radiolabelled by second-strand DNA synthesis and digested with the appropriate restriction enzyme. Restriction fragments were separated by electrophoresis in polyacrylamide gel and quantified using a beta-imager to calculate the percentage of revertants in the sample.

Sequence heterogeneity assay

To identify unstable sites in the viral genome, the virus was serially passaged ten times, its cDNA PCR-amplified and cycle-sequenced. Overexposed X-ray films were inspected for the presence of minor bands indicating possible sequence heterogeneities, and each suspected site was investigated with a specific MAPREC assay [6].

Genetic constructs

A cDNA plasmid containing the Sabin 2 virus genome, obtained from Dr. Andrew Macadam (NIBSC), was found to contain three differences from the Sabin 2 RNA sequence. The plasmid

DNA was corrected using a ligase chain reaction method, and the virus was re-derived by in vitro RNA transcription and RNA transfection into Vero cells. Site-directed point mutations were introduced using full-length PCR method [7].

MNVT

MNVT was performed according to the WHO recommendations [8].

Results

Our work started with the development of MAPREC for the 472-C reversion in type 3 OPV. We found that, although all batches contain some 472-C, those which failed the MNVT contained increased amounts of the reversion [4]. The correlation was so good that it was possible to predict, from the 472-C content, whether a vaccine batch would pass or fail the MNVT (Fig. 1). In several blind studies we correctly identified all unacceptable vaccine lots of type 3 OPV [5]. The WHO then organized an International Collaborative Study to introduce MAPREC to other national control authorities and to vaccine manufacturers. This study, led by CBER/FDA and NIBSC, with the participation of laboratories in Belgium, Canada, France, Germany, Holland, Italy, Japan, Russia, UK, and USA, demonstrated that, even though the method required a certain level of methodological sophistication, each laboratory was able to set up and use the method successfully to discriminate between acceptable and unacceptable batches of vaccine. In the course of the study, CBER/FDA and NIBSC developed the necessary reference reagents, which were validated and accepted by the WHO Expert Committee on Biological Standardization as International References for MAPREC.

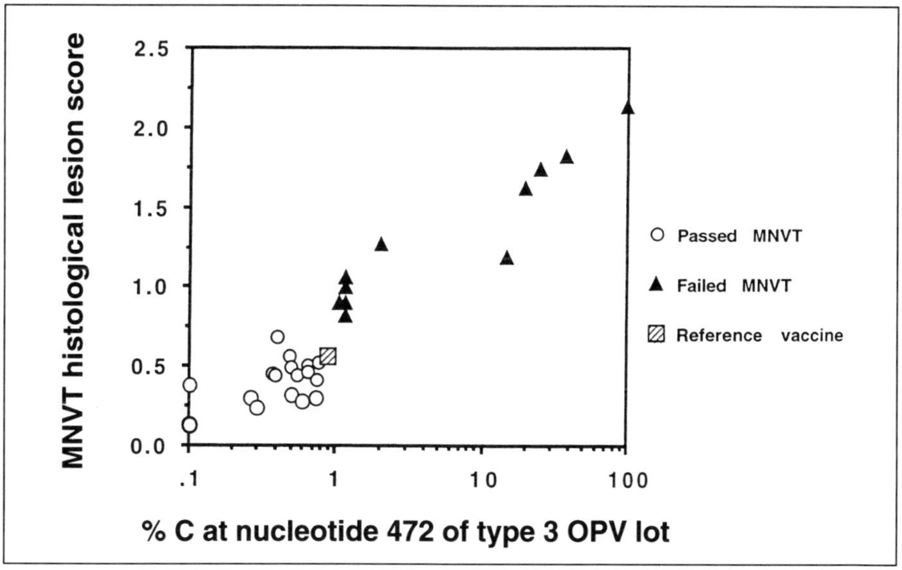

Fig. 1: 472-C revertant content vs. lesion score in monkeys.

An important question is whether there is only one pathway of reversion leading to a loss of attenuated phenotype by a vaccine virus, and, consequently, whether we can rely on testing only one or few nucleotide substitutions to judge the acceptability of vaccine batches. We have performed an extensive screening of genomes of all three types of attenuated polioviruses to identify unstable genomic sites where mutations can occur both in vitro and in vivo. The list of such sites was surprisingly short, and the major unstable sites in type 3 poliovirus were shown to be neutral with respect to attenuation properties ([8], and to be published elsewhere).

Besides enabling laboratories to discriminate between acceptable and unacceptable vaccine batches, MAPREC may reveal information about the history and genetic make-up of each individual batch. During our studies of the genetic stability of the Sabin 3 strain, we identified mutations that are specific for some sub-strains, or are selected upon replication in certain cell cultures [6, 8]. For instance, a C→U mutation at nucleotide 2493 is specific for vaccines made from some strains re-derived from Sabin 3 in Japan, China and Russia; this mutation is present in the Sabin Original substrain, often reaching or exceeding 50%, while in vaccines made from the «Pfizer» strain it rarely reaches 1% [9]. Pfizer-derived vaccines made in Vero cells contain measurable amounts of 4935-G, while vaccines made from the same strain in AGMK cultures do not. Therefore, a mutation profile can tell us a great deal about the different batches of vaccine, including the seed virus used for production, the passage level, and the cell substrate. Although mutations themselves occur as random events, their accumulation is a much more orderly process. With mutation rates reaching 10^{-3}-10^{-4} per nucleotide for each round of viral replication, and the number of live viral particles in each dose of OPV reaching the order of 10^6, it is safe to conclude that all possible point mutations are present in each dose of the vaccine. Content of the mutations, however, remains very low unless a mutation provides virus with replicative advantages. Selectable mutations that accumulate in viral populations are extremely rare, since a vast majority of mutations are deleterious or neutral. Selective pressure depends on replication conditions, and therefore the pattern of mutations that accumulates in a virus stock is both a function and a measure of the conditions of the replicative process. Thus, consistent viral growth conditions are expected to result in a consistent pattern of mutations (mutational profile). The content of selectable mutations can be a relative measure of the number of replication cycles a virus has undergone. Even though the rate of mutant accumulation may vary for each mutation and for each set of culture conditions, in general, sets of selectable mutations tend to accumulate together. We have found that vaccine lots containing excessive amounts of 472-C generally have increased content of other mutations (e.g. 2493-U) as well. Such «sentinel» mutations that do not themselves adversely affect vaccine virus properties and are only a gauge of viral replication can be used effectively to monitor consistency.

In summary, we found that all acceptable type 3 OPV lots contained less than 0.8% 472-C, while batches that failed the MNVT contained more than 1% of these revertants. Determination of the 472-C content allowed us to identify all unacceptable batches. WHO has all the necessary reagents for establishing the routine use of MAPREC for controlling the consistency of type 3 OPV production.

Before this methodology can be applied for monitoring consistency of two other types of OPV, some changes should be made. The reason is that during almost 40 years of OPV testing in monkeys, no batch of OPV 1 or OPV 2 has repeatedly failed the MNVT. An extensive search for such lots revealed only a small number of suspicious batches. These batches were reported to have failed MNVT once, but upon repeated tests in monkeys all such batches were found to be acceptable. Therefore to study revertants in OPV of types 1 and 2 we had to rely on our own experimentally prepared samples. Two types of reversion occurred in the Sabin 1 strain grown in cell cultures: 480–G→A which is a direct back mutation to the wild-type nucleotide and 525–U→C, a compensatory change in the nucleotide complementary to 480 across the stem loop [10]. All batches of OPV 1 contained measurable amounts of both these mutations. We serially passaged Sabin 1 virus and tested samples in monkeys. To our surprise, high passages containing as much as 17% of both 480-A and 525-C (compared to a normal content of 2%) passed the MNVT. One such sample was passaged twice at 37°C, and was still found to be acceptable by the MNVT [11]. These results suggest that the MNVT is not discriminatory enough to identify clearly substandard batches and that MAPREC would be a more appropriate tool for monitoring consistency of OPV 1 production. Thus, unlike type 3 OPV, no loss of attenuation occurred when type 1 OPV was passaged beyond the level used for vaccine production. Therefore any mutations that may accumulate are not predictive of the MNVT results, and can be monitored for consistency purposes only.

A similar situation was found with type 2 OPV. Recently a few batches that initially failed a MNVT were identified by a manufacturer. We used this opportunity to look for mutations that may have accumulated in these batches and caused the failure. To do this, we passaged the batch six times under conditions similar to those used for manufacturing. During passaging, the virus accumulated about 20% of a 480–A→G reversion in the 5'-UTR as well as about 17% of additional mutations in the capsid protein VP1 (3363–A→G and 3364–G→A) -- mutations previously recognized both in vitro and in vivo. Testing samples in the MNVT showed that the original putatively «failed» batch of OPV 2 had only a low level of monkey neurovirulence. In the most revealing experiment, a batch of type 2 OPV passaged six times past the maximum level used for vaccine production nonetheless passed the MNVT. None of the mutations that accumulated in the sample changed its neurovirulence, either alone or when present together. In the same experiment, a genetically engineered construct with 100% 481-G also passed the WHO MNVT, as did a batch of OPV prepared from an alternative seed virus (Chung 2) containing 100% 481-G (Table 1).

Table 1: MNVT results for Sabin 2 samples.

Sample	Mutant content		Lesion score		Test result
	481-G	**VP1 mutations**	**Test sample**	**Reference**	
Lot A	1%	0.34%	0.65 ± 0.51	0.61 ± 0.47	pass
Lot A + 6 passages	20%	17%	0.73 ± 0.31	0.61 ± 0.47	pass
Construct with 100% 481-G	100%	0%	0.46 ± 0.17	0.61 ± 0.47	pass
Lot B (alternative seed)	100%	0%	0.15 ± 0.09	0.68 ± 0.35	pass

Therefore we conclude that genetic stability of both Sabin 1 and Sabin 2 strains of OPV far exceeds the ability of the MNVT to detect any inconsistencies in genetic composition of the vaccine viruses.

Discussion

Unlike the MNVT, MAPREC can easily detect changes in the normal genetic make-up of vaccine viruses and, therefore, appears to be better suited for monitoring vaccine production. Here I would like to propose a new concept for molecular consistency monitoring of live viral vaccines (Fig. 2). In phase one of the establishment of vaccine production, a few initial batches should be produced and examined for the presence of mutations that have a natural propensity to accumulate under production conditions (that is, we should determine the mutational profile of the vaccine). These batches should be concurrently tested by all available biological assays, e.g. MNVT, rct40 marker, d-marker, etc. As soon as the consistency of production has been estab-

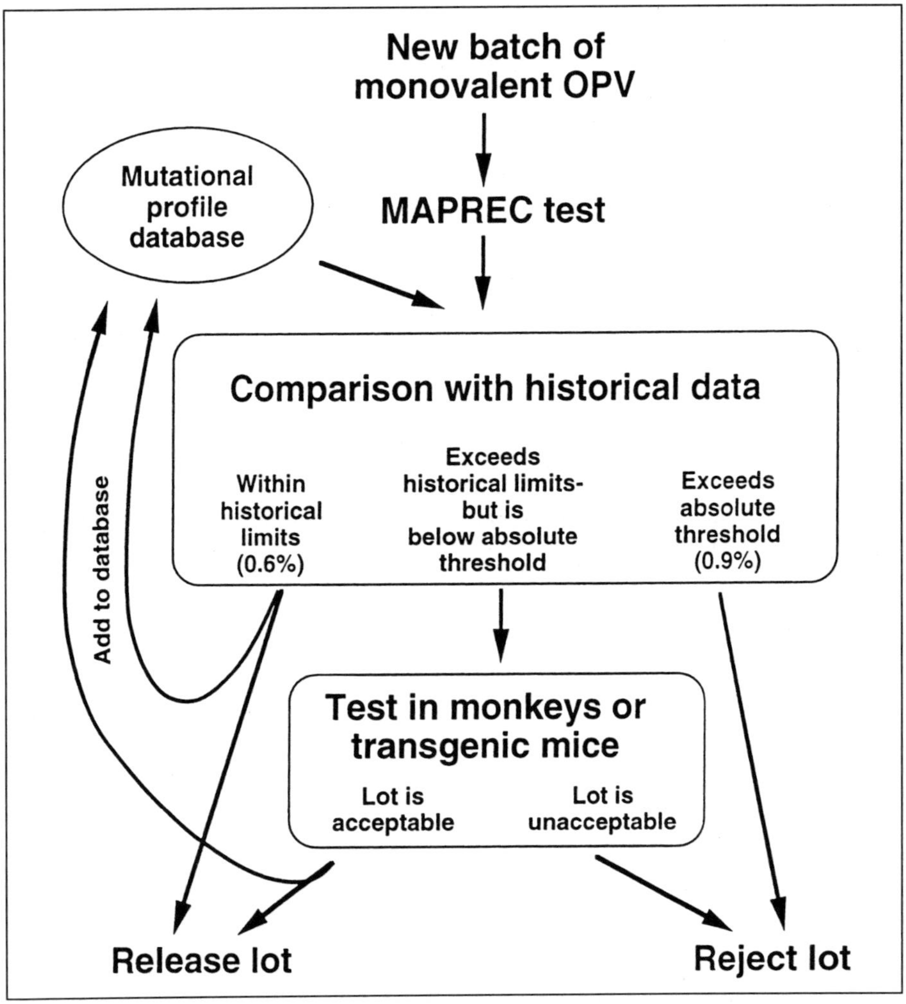

Fig. 2: Consistency approach to acceptability of type 3 OPV.

lished, all further batches can be monitored using only molecular methods, with the aim of maintaining the same mutational profile established during the developmental phase of manufacturing. Should any irregularities in the profile be detected, then MNVT or other relevant biological assay should be used, and troubleshooting initiated to determine what caused the change in the mutational profile.

It is important to stress that this consistency model requires no prior knowledge of the molecular determinants of neurovirulence, but rather relies on «sentinel» mutations that merely gauge the replication history of the virus. This model should be applicable to all vaccines containing live attenuated RNA viruses; its use will advance the safety of viral vaccines since at present there is no consistency test routinely used during production of live viral vaccines other than OPV. Currently, we are working on the development of similar molecular tests for Mumps Live Vaccine (MLV) and Yellow Fever vaccine. Attenuated viruses were passaged under conditions of vaccine production, and analysed using sequence heterogeneity assay. The Yellow Fever virus (17D strain) was found to be extremely stable. Thus far, we have been able to identify only two mutations and are working on developing MAPREC assays for quantifying them. In the Jeryl Lynn® strain of the MLV we have identified a number of changes, and hope to develop a workable molecular consistency assay.

Finally, we want to emphasize that it is both the recent progress in biotechnology and WHO leadership that allowed us to suggest that we may safely shift from the routine use of expensive, cumbersome, and intrinsically variable biological assays to more precise and accurate molecular surrogate methods. As with any surrogate method, molecular testing can only measure the correlates of a biological property, which they do reliably and easily. On the other hand, the MNVT and other biological assays will remain the ultimate safety tests for vaccines. Their use, however, may be limited to the developmental phases of vaccine production and to troubleshooting.

Acknowledgement

I am grateful to Dr. Ghennady Rezapkin, Dr. Li Ju Fan, Dr. Eugenia Dragunsky, Dr. Inessa Levenbook, Dr. David Asher, Ms. Joan Enterline and Ms. Monica Parker, and other present and former members of the Laboratory of Method Development, CBER, FDA, for their contributions to this work.

References

1 Sabin AB: Properties and behavior of orally administered attenuated poliovirus vaccine. Journal of the American Medical Association 1957;164:1216-1223.

2 WHO: Requirements for poliomyelitis vaccine (oral). World Health Organization. 1990.

3 Minor PD, Macadam AJ, Stone DM, Almond JW: Genetic basis of attenuation of the Sabin oral poliovirus vaccines. Biologicals 1993;21:357-363.

4 Chumakov KM, Powers LB, Noonan KE, Roninson IB, Levenbook IS: Correlation between amount of virus with altered nucleotide sequence and the monkey test for acceptability of oral poliovirus vaccine. Proceedings of National Academy of Sciences of USA 1991;88:199-203.

5 Chumakov K, Norwood L, Parker M, Dragunsky E, Taffs R, Ran Y, Ridge J, Levenbook I: Assessment of the viral RNA sequence heterogeneity for control of OPV neurovirulence, in

Brown F, Lewis BP (eds): Poliovirus Attenuation: Molecular Mechanisms and Practical Aspects. Dev Biol Stand 1993, vol 78, pp 79-89; discussion pp 88-89.

6 Rezapkin GV, Chumakov KM, Lu Z, Ran Y, Dragunsky EM, Levenbook IS: Microevolution of Sabin 1 strain in vitro and genetic stability of oral poliovirus vaccine. Virology 1994;202:370-378.

7 Chumakov KM: PCR Engineering Of Viral Quasi-Species – a New Method to Preserve and Manipulate Genetic Diversity Of RNA Virus Populations. Journal of Virology 1996;70:7331-7334.

8 Rezapkin GV, Norwood LP, Taffs RE, Dragunsky EM, Levenbook IS, Chumakov KM: Microevolution of type 3 Sabin strain of poliovirus in cell cultures and its implications for oral poliovirus vaccine quality control. Virology 1995;211:377-384.

9 Chumakov KM, Norwood LP, Parker ML, Dragunsky EM, Ran YX, Levenbook IS: RNA sequence variants in live poliovirus vaccine and their relation to neurovirulence. Journal of Virology 1992;66:966-970.

10 Kawamura N, Kohara M, Abe S, Komatsu T, Tago K, Arita M, Nomoto A: Determinants in the 5' noncoding region of poliovirus Sabin 1 RNA that influence the attenuation phenotype. J Virol 1989; 63:1302-9.

11 Rezapkin GV, Alexander W, Dragunsky E, Parker M, Pomeroy K, Asher DM, Chumakov KM: Genetic stability of Sabin 1 strain of poliovirus: Implications for quality control of Oral Poliovirus Vaccine. Virology 1998;245:183-187.

Dr. K.M. Chumakov, Center for Biologics Evaluation and Research, Food and Drug Administration, 1401 Rockville Pike, HFM-470, Rockville, MD 20852, USA.

Brown F, Griffiths E, Horaud F, Schild GC (eds): A Celebration of 50 Years of Progress in Biological Standardization and Control at WHO.
Dev Biol Stand. Basel, Karger, 1999, vol 100, pp 75-82

Alternatives to the Use of Animals for Bacterial Toxins and Antitoxins

D. Sesardic

Division of Bacteriology, National Institute for Biological Standards and Control, Potters Bar, Hertfordshire, U.K.

Key words: Therapeutic agents, bacterial toxins, antitoxins and vaccines, alternative assays, zinc-dependent endopeptidase assay.

Abstract: The potency of several novel botulinum toxin-derived biological therapeutic products now in routine medical use is determined exclusively by in vivo methods. In addition, large numbers of animals continue to be used for the potency and safety testing of therapeutic antitoxins and toxoid vaccines. There is thus an increasing need to develop acceptable alternative assays for toxicity testing of clostridia neurotoxins which could be applied to different toxin derived therapeutic agents, including vaccines. Scientific advances in the understanding of the mode of action of clostridial neurotoxins have now provided the basis for improving conventional testing procedures.

INTRODUCTION

Biological substances used in medicine are usually chemically complex, often poorly defined and may contain many non-functional components, including degradation products. Thus, in controlling such agents it is important to assess their biological activity [1]. Currently, those assays providing the specificity and sensitivity necessary for the testing of biological therapeutics are largely reliant on in vivo systems.

Bacterial exotoxins are the causative agents of a number of life threatening diseases. Amongst these, diphtheria and tetanus would be responsible for major loss of life if it were not for the success of vaccination against them. Whilst botulism is not as prevalent, it is life threatening and the aetiological factor, botulinum toxin, is being increasingly used as a therapeutic agent because of its selective pharmacological property as a potent neuromuscular paralytic agent [2]. Current methods for estimating the potency of bacterial toxoid vaccines and antitoxins [3] and of botulinum toxin type A (BoNT/A) [4] require large numbers of animals. Whilst efforts have been made to reduce the number and refine the biological end points used in potency testing of these therapeutics [5], the ultimate aim is to develop methods which would enable the entire replacement of current in vivo procedures.

Clostridium botulinum toxins are potent neuroparalytic agents which act presynaptically at the neuromuscular junction, their primary site of action, where they induce flaccid paralysis by blockade of acetylcholine release. The structurally related tetanus toxin (TeNT) causes a spastic paralysis due to its retrograde transport to the spinal cord where it blocks the release of neurotransmitters. Both botulinum and tetanus neurotoxins target one or more of a set of proteins essential for neurotransmitter release from the synaptic vesicle [6]. This proteolyttic activity, confirmed as zinc-dependent endopeptidase activity [7], is highly specific as each toxin is selective for a single bond within their target protein involved in synaptic exocytosis [8].

Using this information it has been possible to develop in vitro biochemical assays for those clostridial neurotoxins which are used therapeutically as modifiers of neural activity (BoNT/A and BoNT/B) or as immunogens (TeNT), with the aim of reducing the number of animals used in quality control.

In Vitro Assays for Clostridia Toxins

During the last several years, a number of in vitro assays have been developed to detect and measure BoNT activity [6]. Many of the strategies developed are based on immunodetection of toxins. These are not suitable when it is necessary to measure biological activity of the toxin [9]. This is because immunoassays have limited ability to distinguish between biologically active toxin and toxin which is inactive. Because of such limitations, immunoassays could never serve as acceptable alternatives for the potency or safety testing of biologicals derived from bacterial toxins where it is absolutely essential to quantify the active component [10].

There have been important advances in understanding the mode of action of several bacterial toxins over the last ten years. Most important of these was the discovery that the light chain of neurotoxins possess zinc-dependent endopeptidase activity and confirmation that neurotoxins act on nerve terminals via an enzymatic process [6]. As each neurotoxin cleaves a specific bond within their target protein, this is a highly selective mechanism of action. Tetanus and botulinum type B toxins cleave vesicle-associated membrane protein VAMP-2, also known as synaptobrevin, at residues Gln_{76}-Phe_{77} [11]. Botulinum toxin type A cleaves a 25-kDa synaptosomal-associated protein known as SNAP-25, at residues Gln_{197}-Arg_{198} and type E toxin selectively cleaves the same protein at residues Arg_{180}-Ile_{1181} [12]. Figure 1 shows the crystallographic structure of synaptic fusion complex proteins with the location of BoNT/A and BoNT/E and BoNT/B and TeTN cleavage sites on SNAP-25 and VAMP-2 respectively.

The enzymatic property of these neurotoxins has now been exploited to measure toxin activity and as the basis for the development of in vitro assays. Several approaches for detecting the cleavage reaction have been investigated. Studies with synthetic VAMP-2 and BoNT/B toxin confirmed that it was possible to use a synthetic substrate to demonstrate toxin endopeptidase activity in vitro and to use RP-HPLC to monitor the reaction [13]. A similar approach was adopted for BoNT/A using synthetic SNAP-25 [14, 15]. A fluorescent substrate was used for TeTN detection [16]. In this assay, the cleavage of a fluorescent synthetic peptide fragment of VAMP-2 containing the fluorescent Pya^{88} moiety was separated by RP-HPLC and the cleavage product detected by fluorescence emission at 377 nm. An alternative approach included the use of capillary electrophoresis, which has high resolving power and the ability to separate the substrate from the product, to detect neuro-

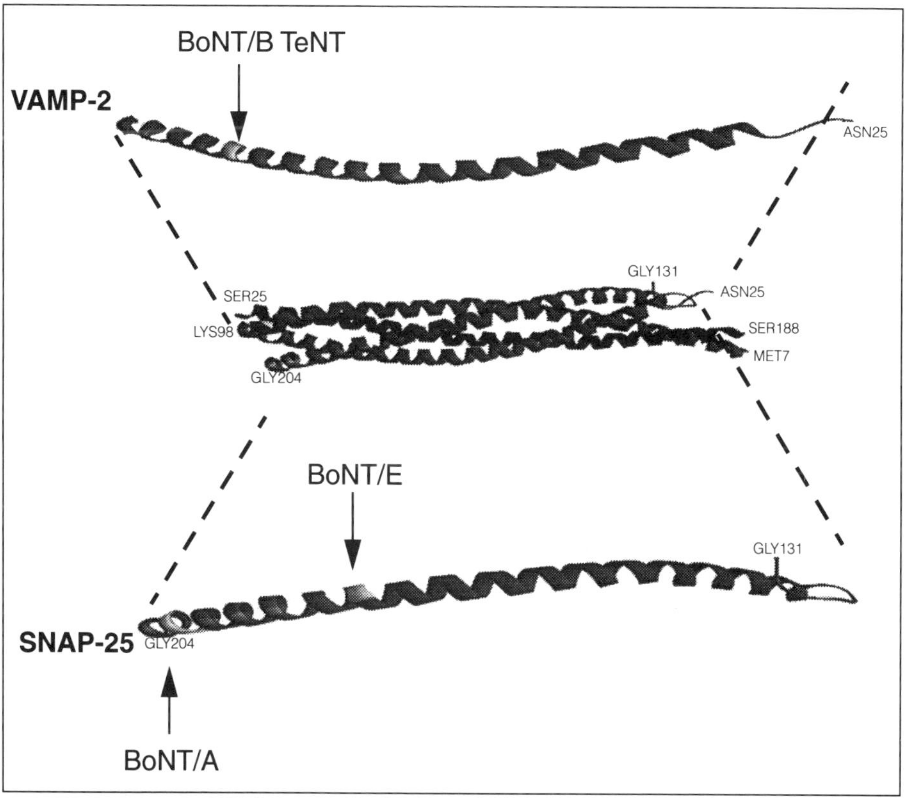

Fig. 1: Backbone ribbon drawing of the synaptic fusion complex proteins involved in synaptic exocytosis as predicted from the crystal structure [8]. Location of BoNT/A and BoNT/E and of BoNT/B and TeTN cleavage sites are shown on SNAP-25 and VAMP-2 proteins respectively.

toxin mediated proteolysis of SNAP-25 [17]. All these approaches were satisfactory for the detection of toxin proteolytic activity but lacked the sensitivity required for detection of the small quantities of active toxin present in therapeutic formulations. The most sensitive in vitro endopeptidase assay was achieved by using targetted antibodies for immunodetection of the neurotoxin cleaved substrates. Microtitreplate endopeptidase assays have now been reported for BoNT/B [18] and BoNT/A [19]. Using a recombinant peptide of SNAP-25 (aa residues 134-206) we have shown that the sensitivity of the in vitro endopeptidase assay was better that that of the mouse assay in detecting the active BoNT/A in therapeutic preparations. Furthermore, we have demonstrated an excellent correlation with results using the in vivo bioassay. [19]. Since the development of the endopeptidase assay, batches of therapeutic BoNT/A have been tested in parallel with the in vivo bioassay for batch release testing. As indicated in Figure 2, comparable potency estimates were obtained when using the in vitro and in vivo assays.

Although the in vitro endopeptidase assay has considerable potential to replace the in vivo mouse assay for batch release potency testing of botulinum toxins used

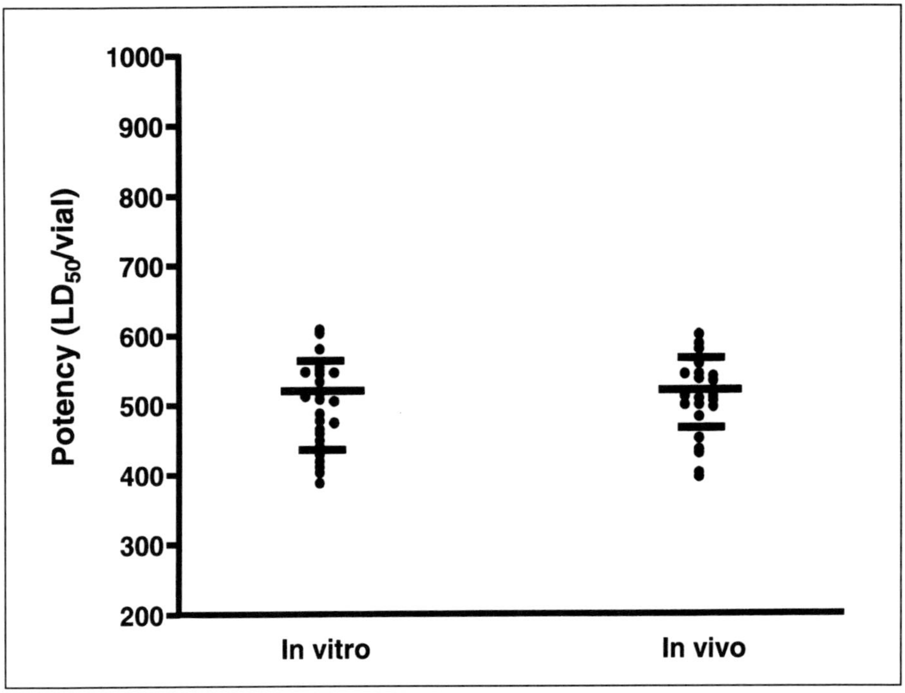

Fig. 2: Comparison of BoNT/A activity determined by in vivo mouse assay and in vitro endopeptidase assay for 26 different batches of therapeutic preparation of botulinum toxin type A obtained from one manufacturer for batch release testing. Both activities, expressed as LD50 per vial, were relative to the same in-house standard.

therapeutically, it must be emphasised that the factors contributing to the biological and pharmacological activity of botulinum toxin may not be dependent on intracellular mode of action alone. The validity of the endopeptidase assay described above for assessing the biological potency of botulinum toxin will depend upon whether substrate clea-vage is the rate limiting step in the intoxication process. It is worth noting that a recent report on diphtheria toxin, which is functionally related to botulinum toxin in its possession of the receptor binding, internalisation and catalytic domains, has confirmed that intracellular enzymatic activity is the rate-limiting step in the intoxication by diphtheria toxin [20]. Nevertheless, direct measurement of the enzymatic activity of toxin is unlikely to enable the accurate prediction of its therapeutic activity but the same limitations also apply to the existing whole animal bioassay models [6].

Tetanus vaccine is composed of chemically denatured tetanus toxin. Thus, safety testing requires confirmation of freedom from residual toxicity in vivo [3]. As the L-chain of TeNT is a zinc-endopeptidase, which cleaves vesicle associated protein (VAMP-2), we have explored the utility of synthetic substrates, containing the TeNT cleavage site, for an in vitro endopeptidase assay for this toxin. Preliminary results are encouraging. Using both RP-HPLC and immunodetection we were able to demonstrate in vitro endopeptidase activity with crude formulations of TeNT, with a sensitivity close to that of the mouse bioassay [21, 22]. However, further validation studies will be required to confirm the usefulness of such an assay in the safety testing of tetanus vaccines.

In Vitro Assays for Clostridia Antitoxins

Following immunisation with vaccines against bacterial toxins (toxoid vaccines) antibodies against multiple epitopes are produced. Only some of these will be neutralising. The ability of toxoid vaccines to induce toxin neutralising antibodies has provided the basis for the use of therapeutic antitoxins and immunoglobulins for the treatment of diseases caused by neurotoxins secreted by strains of *Clostridium botulinum*. Currently, the potency of therapeutic antitoxins is assayed exclusively using in vivo bioassays [3]. Immunoassays are widely used for the detection of toxin specific antibodies in serum samples both from vaccinated individuals, as predictors of population immunity and from vaccinated animals, as indicators of vaccine immunogenicity. Current immunoassays however, cannot distinguish between neutralising and non-neutralising antibodies, although for many antigens, where the neutralising epitope may also be immunodominant, good correlation between immunodetection and protective potency can be observed [10]. However, assays that can selectively discriminate between toxin neutralising antibodies are desirable and are essential for the potency testing of therapeutic antitoxin and immunoglobulin formulations.

Whilst the biological activity of clostridial neurotoxins depends upon the integrity of all three of the functional domains, neutralising anti-toxin activity appears to be directed against the receptor binding domain [23]. Protective responses to botulinum and tetanus toxins have been reported following immunisation with only toxin heavy chain containing the receptor binding property. However, a monoclonal antibody with partial ability to block BoNT/A activity in vivo has been mapped to the toxin light (L) chain containing the enzymatic activity of the toxin [24]. We have therefore investigated the use of the toxin endopeptidase assay for the in vitro detection of toxin neutralising antibodies. We have demonstrated that BoNT/A in vitro endopeptidase activity is specifically inhibited by type A toxin antiserum but not by antiserum to other neurotoxins, suggesting that the assay can also be used to detect selectively antibodies to the catalytic domain of the toxin [25]. Using purified BoNT/A toxin and equine International Standard for type A antitoxin (500 IU/vial) it was possible to detect 0.002 IU/ml of type A antitoxin activity, as the lowest concentration on antitoxin to inhibit BoNT/A toxin proteolytic activity. Three different preparations of equine antitoxin (monovalent, trivalent and botulinum F $(ab')^2$ heptavalent) and human botulism immune globulin intravenous, with a wide range of potency (25.0-3,000 IU/ml) were tested by in vivo toxin neutralisation and by in vitro assay. The results in Figure 3 indicate that the in vitro assay provided comparable information to the bioassay for quantification of toxin neutralising antibodies.

We have developed a new in vitro assay that is capable of detecting selectively antibodies to the catalytic domain of the toxin. It may provide an alternative to in vivo assay for monitoring the potency of therapeutic antitoxins. The assay may not have the ability to detect all of the antibodies in vaccinated individuals that are responsible for the protective response in vivo, as many reports have identified the importance of the receptor binding domain in protection, but further studies, particularly with recombinant and chimerical toxins (mixing heavy and

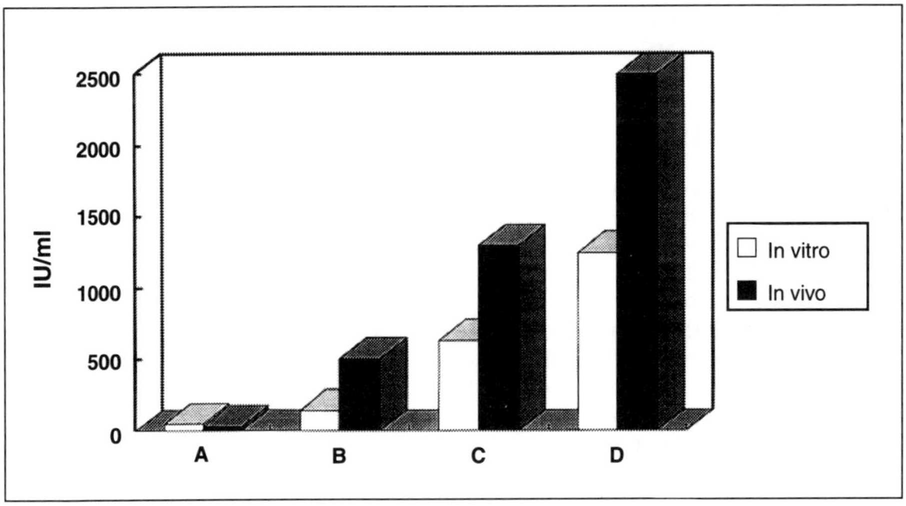

Fig. 3: Comparison of potency determined by in vivo toxin neutralisation assay (shaded bars) and by in vitro endopeptidase assay (open bars) for four different therapeutic formulations containing type A antitoxin. Both activities were expressed in IU/ml relative to the International Standard for type A antitoxin.

light chain domains from different serotypes), will be important. The assay has the potential for the rapid in vitro testing of human sera in patients after therapy with BoNT/A to identify potential non-responders due to the presence of toxin-specific antibodies.

Conclusions

Valid alternative assays for estimating the efficacy and safety of biological products must be based on end points comparable to those responsible for their in vivo activity. Information derived from studies on the structure, mode of action and immunogenicity of bacterial toxins has provided the basis for the development of alternative bioassays, more suitable than those previously proposed, for testing the efficacy and safety of therapeutic products derived from bacterial toxins. This applies to toxins, antitoxins and vaccines, as the protective efficacy of vaccines depends on the level of toxin neutralising antibodies induced. Much progress has been made in the development of in vitro enzymatic assays for some clostridial toxins used therapeutically. The assay for BoNT/A toxin developed at NIBSC has been used successfully for monitoring the potency of this toxin in clinical formulations. Wider use of this assay should result in significant reduction in the use of animals for this purpose. A similar enzymatic assay has been developed for tetanus toxin which could be adopted for the specific and residual toxicity testing of tetanus vaccine. Application of this assay to quality control of toxoid vaccines could entirely replace the use of animals for this purpose. However, each application of the assays described will still require careful validation studies as no one assay so far developed has the capacity to detect all biological properties of bacterial toxins.

REFERENCES

1 Jeffcoate SL, Corbel MJ, Minor PD, Gaines-Das RE, Schild GS: The control and standardisation of biological medicines. Proc Royal Soc Edinburgh 1993;101B:207-226.

2 Jankovic J, Brin MF: Therapeutic uses of botulinum toxin. N Engl J Med 1991;324:1186-1194.

3 British Pharmacopoeia 1998. Volume II.

4 McLellan K, Gaines Das R, Ekong TAN, Sesardic D: Therapeutic Botulinum type A toxin: factors affecting potency. Toxicon 1996;34:975-985.

5 Sesardic D, McLellan K, Ekong TAN, Gaines Das R: Refinement and validation of an alternative bioassay for potency testing of therapeutic botulinum type A toxin. Pharmacology & Toxicology 1996:78;283-288.

6 Pearce LB, First ER MacCallum RD, Gupta A: Review Article. Pharmacologic characterization of botulinum toxin for basic science and medicine. Toxicon 1997;35:1373-1412.

7 de Paiva A, Ashton AC, Foran F, Schiavo G, Montecucco C, Dolly JO: Botulinum A like type B and tetanus toxins fulfils the criteria for being a zink-dependent protease. J Neurochem 1993; 61:2338-2341.

8 Sutton RB, Fasshauer D, Jahn R, Brunger AT: Crystal structure of a SNARE complex involved in synaptic exocytosis at 2.4 Å resolution. Nature 1998;395:347-353.

9 Ekong TAN, McLellan K, Sesardic D: Immunological detection of Clostridium botulinum toxin type A in therapeutic preparations. J Imm Methods 1995;180:181-191.

10 Sesardic D: Requirements for valid alternative assays for testing of biological therapeutic agents, in Brown F, Cussler K, Hendriksen C (eds): Replacement, Reduction and Refinement of Animal Experiments in the Development and Control of Biological Products. Dev Biol Stand. Basel, Karger, 1996, vol 86, pp 311-318.

11 Schiavo G, Poulain B, Benfenati F, DasGupta BR, Montecucco C: Novel targets and catalytic activities of bacterial protein toxins. Trends Microbiol 1993;1:170-174.

12 Blasi J, Chapman ER, Link E, Binz T, Yamasaki S, DeCamilli P, Sudof TC, Niemann H, Jahn R: Botulinum neurotoxin A selectively cleaves the synaptic protein SNAP-25. Nature 1993;365: 160-163.

13 Shone CC, Quinn CP, Wait R, Hallis B, Fooks SG, Hambleton P: Proteolytic cleavage of synthetic fragments of VAMP isoform-2 by botulinum type B neurotoxin. Eur J Biochem 1993; 217:965-971.

14 Schmidt JJ, Bostian KA: Endoproteinase activity of type A botulinum neurotoxin: substrate requirements and activation by serum albumin. J Prot Chem 1997;16:19-26.

15 Sesardic D, Corran PH, McLellan K, Feavers IM, Sesardic D: In vitro assays for estimating the activity of therapeutic preparations of botulinum toxin. Animal Alternatives, Welfare and Ethics 1997;27:1033-1038.

16 Soleilhac J-M, Cornille F, Martin L, Lenoir C, Fournie-Zaluski M-C, Roques BP: A sensitive and rapid fluorescence-based assay for determination of tetanus toxin peptidase activity. Anal Biochem 1996;241:120-127.

17 Corran PH, Sesardic D: A capillary electrophoresis-based assay of botulinum Type A toxin proteolytic activity. Poster Abstract. Presented at: Biomedical Aspects of Clostridial Neurotoxins International Conference Oxford, July 1996, Oxford UK.

18 Hallis B, James BAF, Shone CC: Development of novel assays for botulinum type A and B neurotoxins based on their endopeptidase activities. J Clin Microbiol 1996:1934-1938.

19 Ekong TAN, Feavers IM, Sesardic D: Recombinant SNAP-25 is an effective substrate for Clostridium botulinum type A toxin endopeptidase activity in vitro. Microbiology 1997;143:3337-3347.

20 Falnes PO, Olsnes S: Modulation of the intracellular stability and toxicity of diphtheria toxin through degradation by the N-end rule pathway. EMBO J 1998;17:615-625.

21 Corran P, Ekong TAN, Gee C, Sesardic D: HPLC-based methods for the measurement of clostridial neurotoxin proteolytic activity. Poster presented on: Alternatives to animals in the

development and control of biological products for human and veterinary use. 24-26 September 1998, London, UK, p 67.

22 Ekong TAN, Sesardic D: Microtitre-plate endopeptidase assay for tetanus toxin. Poster presented on: Alternatives to animals in the development and control of biological products for human and veterinary use. 24-26 September 1998, London, UK, p 71.

23 Clayton MA, Clayton JM, Brown DR, Middlebrook JL: Protective vaccination with a recombinant fragment of Clostridium botulinum neutotoxin serotype A expressed from synthetic gene in Escherichia coli. Infec Immun 1995;63:2738-2742.

24 Hallis B, Fooks S, Shone CC, Hambleton P: Characterization of monoclonal antibodies to botulinum type A neurotoxin, in Das Gupta BR (ed.): Botulinum and Tetanus Neurotoxins. New York, Plenum Press, 1993, pp 433-436.

25 Martin H, Sesardic D: Testing of neutralising antibodies to botulinum A toxin: implications for potency testing of therapeutic antitoxins. Abstract presented at: European Clostridia Conference, 4-7 October 1998, Teistungen, Germany.

Dr. D. Sesardic, Division of Bacteriology, National Institute for Biological Standards and Control, Blanche Lane, South Mimms, Potters Bar, Hertfordshire EN6 3QG, U.K.
E-mail: dsesardic@nibsc.ac.uk

Brown F, Griffiths E, Horaud F, Schild GC (eds): A Celebration of 50 Years of Progress in Biological Standardization and Control at WHO.
Dev Biol Stand. Basel, Karger, 1999, vol 100, pp 83-93

Cytokine and Growth Factor Standardization: the Need for Research to Answer Modern Issues

A.R. Mire-Sluis

Division of Immunobiology, National Institute for Biological Standards and Control, Potters Bar, Herts., U.K.

Key words: Cytokines, growth factors, bioassays, standards.

Abstract: New and novel cytokines are being discovered, cloned and entered into clinical trials at such a rate that often the biological activities of these proteins are poorly understood during their development as therapeutic agents. When estimating the biological activity of different preparations with different specific activities by bioassay, mass units cannot be used. Biological activity is therefore expressed as «biological potency units». Due to the pleiotropic nature of cytokine activity, the biological unit requires definition by a standard that is assay-independent. The need to investigate the biological activity of any proposed standard material in a variety of assays, with statistically valid numbers of participants and with appropriate scientific rationale, has led to the design of international collaborative studies that have provided extremely useful data about the biological activity of cytokines. Such information has had a great impact on the choice of material to serve as a standard and has prevented the establishment of inappropriate preparations. Over the years, WHO international standards have been used to reduce the variation in estimates of cytokine preparations within and between laboratories for both immunoassays and bioassays.

INTRODUCTION

Cytokines are proteins that control the maintenance and function of the haematopoietic/immune system [1]. Growth factors are a family of proteins that act mainly on non-haematopoietic tissues [2]. However, there is a degree of overlapping activities between these two broad groups of proteins and for the purpose of this paper, the term cytokines will include growth factors.

More than 150 cytokines have been identified and shown to interact in a complex network to fulfil their biological functions. Since cytokines are involved in many physiological processes it is not surprising that they are also involved in both the pathogenesis and treatment of many diseases [3]. A range of cytokines has been successfully developed as therapeutic agents, such as granulocyte colony stimulating factor (G-CSF) [4], erythropoietin [5] and interferons alpha and beta (IFN-α and IFN-β) [6] and on average, each accounts for 10-50 million doses a year. The discovery that cytokines act as mediators of disease pathology has led to a vast amount of research investigating their role in disease and also to the possibility of using levels of cytokines as diagnostic and prognostic markers. This requires precise and accurate estimates of levels of cytokines in various biological fluids by different laboratories worldwide [7]. If data are to be compared between such laboratories, their results must be expressed in the same way. The use of WHO international reference preparations for designating the activity of biological preparations used in prophylaxis, therapy, or diagnosis helps to ensure the comparability of the activity of these substances worldwide, as well as the reliability of diagnostic measures (Table 1).

Table 1: Cytokine and growth factor international collaborative studies.

Cytokine	Number of participants	Year established
Interferon-alpha, leucocyte	5	1978
Interferon-alpha lymphoblastoid	8	1984
Interferon-alpha 2a	8	1984
Interferon-beta	8	1987
Interferon-alpha D	8	1987
Interferon-alpha 2b	8	1987
Interleukin-2	18	1988
Interleukin-1 alpha	11	1990
Interleukin-1 beta	11	1990
Granulocyte-macrophage colony stimulating factor	29	1992
Granulocyte-colony stimulating factor	29	1992
Macrophage colony stimulating factor	25	1992
Tumour necrosis factor alpha/beta	20	1992
Interleukin-6	12	1993
Basic fibroblast growth Factor	11	1993
Interferon-gamma	13	1994
Insulin like growth factor-1	13	1994
Epidermal growth factor	12	1994
Interleukin-3	53	1994
Interleukin-4	53	1994
Interleukin-8	30	1995

Table 1: Cytokine and growth factor international collaborative studies.

Cytokine	Number of participants	Year established
Platelet derived growth factor-BB	7	1997
Hepatocyte growth factor/scatter factor	11	Submitted 1999
Insulin like growth factor-1	10	Submitted 1999
Stem cell factor	20	Study in progress 1998
Interferon-alpha	97	Submitted 1999
Leptin	9	Submitted 1999

The Need for Cytokine Standards

Assessing the biological activity of cytokines can only be achieved by bioassay at present, where bioassays are defined as an analytical procedure measuring a biological activity of a test substance based on a specific, functional, biological response of a test system.

Various methods of production of cytokine proteins exist and can use cell sources such as E. coli, yeast and Chinese hamster ovary (CHO) cells. The use of the same cell line can result in a different product if grown under different conditions, and manufacturers who use the same cell production source often use different methods to refold and stabilise the protein. This has led to the availability of a number of different forms of each cytokine possessing distinct biological specific activities. This results in a situation where material from one manufacturer can have a very different biological potency than that from another and therefore it is not possible to calibrate correctly such proteins by mass [8].

Whilst the value of accurate estimation of cytokine activity is evident for laboratory researchers, it is even more imperative that manufacturers of cytokines intended for clinical use are able to produce both precise and accurate estimates of cytokine bioactivity. This is a necessity throughout the production process, for batch to batch consistency and during the shelf life of the product to demonstrate stability of the preparation. Measurement of biologically active material is also of importance during pharmacological studies. In addition, the therapeutic window of many cytokines is quite small, particularly for cytokines that have life-threatening side effects such as interleukin-2 and tumour necrosis factor [9].

Information supplied to clinicians should include the specific activity of the product as this can vary considerably between manufacturers. The absence of an internationally agreed biological standard leads to the development of individual in-house standards calibrated in different ways. WHO international collaborative studies have shown that laboratories have developed in-house standards calibrated using mass, concentration (mols or mass/ml) or potency unitages based on either bioassay ED_{50} or other biological or immunological responses. This introduces substantial variation in the estimates of cytokine levels between different laboratories in both bioassays and immunoassays [7].

Design of WHO International Collaborative Studies

A WHO international cytokine standard serves as the primary reference preparation in terms of which all other reference standards should be calibrated. It is necessary to ensure that the material contained in the international standard is suitable as a bioassay standard. Cytokine material can come from various sources and at different stages of development, from research laboratory grade to manufacturers clinical grade material. It is therefore vital to carry out international collaborative studies on a range of materials from different sources (e.g. E. coli, yeast, CHO etc.) as it is impossible to predict which material is most suitable.

It must be noted that clinical grade material does not always serve the purposes of a standard simply because it is produced under good manufacturing practice and is either highly purified or characterised. Therefore, several candidate materials should be included in any collaborative study. Since it is known that cytokines can each exhibit a variety of different biological activities, it is crucial that any material serving as an international standard is shown to perform appropriately in as many different biological and immunological assays as possible and enough laboratories are recruited to achieve this objective. It is also important that the laboratories undertaking cytokine assays are representative of different backgrounds such as research laboratories, national control authorities, hospital laboratories and the biotechnology industry. It is also important that a number of different countries participates.

Participants are sent coded ampoules of lyophilised material to be assayed using their current in-house bioassay or immunoassay method. Within the coded set of ampoules are included two identical ampoules with different codes. This permits accurate assessment of inter- and intra-assay variation when analysing the data provided.

Participants are requested to provide the raw data from each assay performed in addition to information about: (i) the methodology; (ii) the dilution series; (iii) the assay format and the row data provided should be analysed by a single laboratory. This has been shown to be the most appropriate procedure for analysis of a large study since the use of different methods of analysis by different participants can be a source of variation.

The choice of international standard is based on many factors. The material must have been shown to reduce inter-laboratory variability to a relevant extent in as many different bioassays and immunoassays as possible. The standard must be shown to be stable under general laboratory and storage conditions. This requires careful consideration in the design of a collaborative study, where materials assayed from freshly reconstituted ampoules must be distinguished from frozen material. Differences between the relative activities of preparations included in the study amongst different bioassays suggests a need to investigate whether the materials are truly comparable or require individual standards [9].

The need to include enough scientific design into collaborative studies to address these issues has been illustrated in several previous international collaborative studies. These are used as examples of the types of data that such studies can provide about the basic biology of cytokines and how this influences the choice of standard [10].

Case Study 1: The International Collaborative Study of Granulocyte Colony Stimulating Factor

Granulocyte colony stimulating factor (G-CSF) is a 18.5kDa protein that stimulates the production of white cells from bone marrow progenitors and can mobilise stem cells into the peripheral blood. G-CSF is used routinely in the reconstitution of the immune system post bone marrow transplant. There are basically two commercially available G-CSF products – E. coli non-glycosylated and CHO glycosylated forms. Both these preparations were included in the WHO collaborative study, in addition to a research material that was glycosylated and had a single cysteine residue replaced by a serine. Before the study, there had been a great deal of controversy over the relative activities of the clinical products and the role of glycosylation on the biological activity of the protein.

The results of the collaborative study showed that there was a discrepancy between the relative activities of the E. coli non-glycosylated material and the CHO glycosylated material when using either cell line or bone marrow based bioassays [11]. E. coli material appeared to have a much lower specific activity in bone marrow assays than in cell line assays and relatively lower activity when compared to CHO material (Fig. 1A). This difference accounted for the confusion between reports of the activities of G-CSF, where different bioassays had been used. Further investigation has shown that this difference in the results is due to the length of time of the assay, with bone marrow assays taking five to seven days to complete in comparison to one day for cell line assays. Since both the assays are carried out at 37°C, it was the stability of the protein molecule that accounted for most of the differences seen. At 37°C, E. coli material rapidly loses its biological activity, while the glycosylated CHO material appears to be more resistant to heat degradation (Fig. 1B). Therefore, in the longer bone marrow assay, E. coli material loses more activity relative to CHO than in one day cell line bioassays, where heat degradation is minimal. Subtle differences were seen in relative activities even between one and two day cell line bioassays, leading to recommendations about their design. It was vital therefore, that the collaborative study included these different assays, even though cell line assays were gradually replacing bone marrow assays in the measurement of G-CSF activity.

In addition, the research grade material was the only preparation that was not grossly affected by heat degradation. This material performed appropriately in all assay systems and was chosen as the most suitable preparation to serve as the international standard. In fact, without this highly stable preparation to serve as a reference, the controversy over the activities of the clinical products would have been much harder to resolve.

Case Study 2: The International Collaborative Study of Interleukin-8

Interleukin-8 (IL-8) is a member of the chemokine family and is responsible for the chemotaxis of neutrophils and their activation in sites of inflammation. The measurement of levels of IL-8 and understanding its biology are vital to the development of anti-inflammatory agents. IL-8 is a 72 amino acid non-glycosylated protein and its

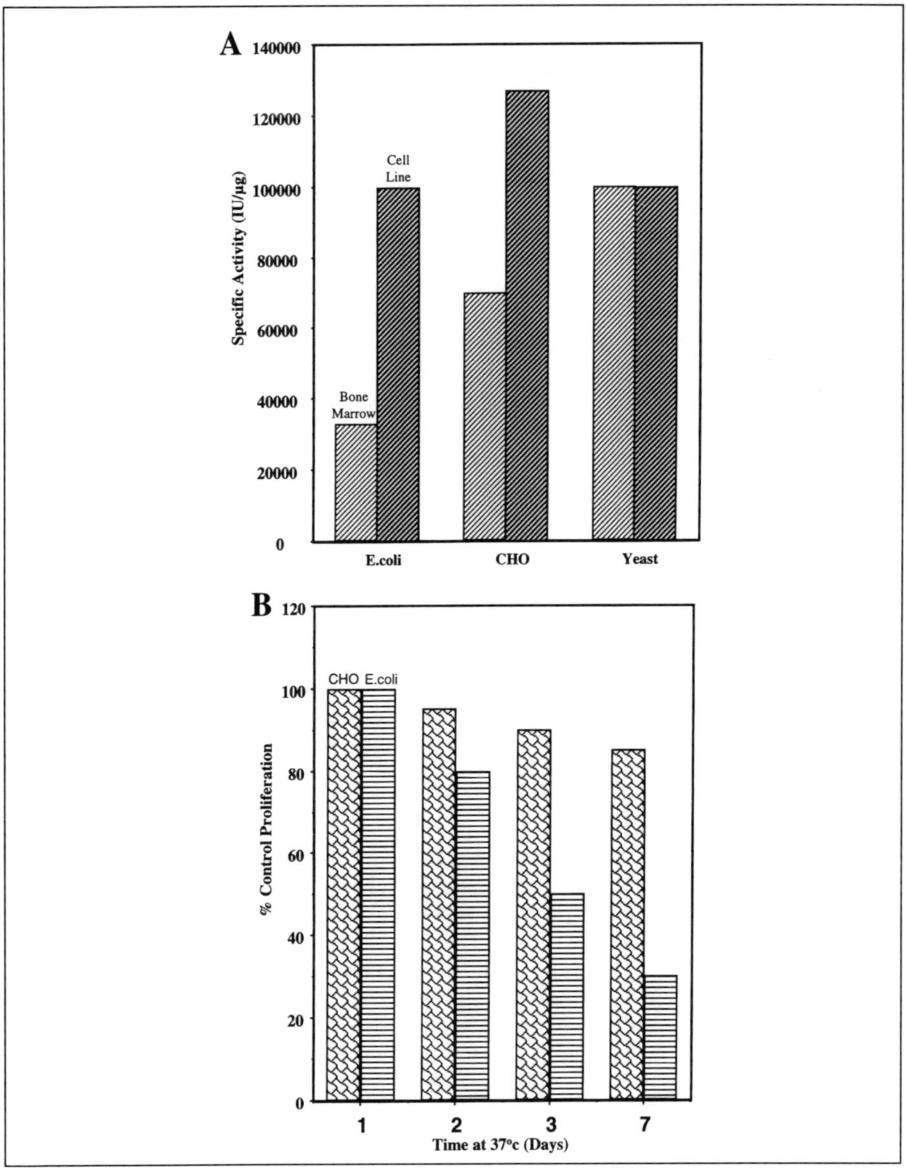

Fig. 1: Results from and succeeding the International Collaborative Study for Granulocyte Colony Stimulating Factor.

Panel A. Three different preparations of G-CSF from the cell sources shown were assayed on either cell line or bone marrow based bioassays as indicated, illustrating the lower and variable specific activity of E. coli derived material compared to CHO material between the two bioassay systems.

Panel B. CHO and E. coli derived G-CSF were incubated at 37°C for the times indicated before bioassay on the GNFS-60 cell line, indicating the increased susceptibility of E. coli derived material to heat degradation.

structure/function relationships are generally unknown. Six preparations of IL-8 were available at the time of the international collaborative study [12]. They were all prepared in E. coli and said to be "identical". However, there had been a great deal of discussion in research reports about the biological activity of IL-8, whether it acted as a dimer or monomer, and concerns over its stability and performance in different biological assay systems. It was therefore decided that all six preparations needed to be included in the study, as each had different relative activities and potentially different spectra of activity. In addition, careful consideration had to be made over the stability of the protein as there were indications that chemokines were prone to unexplained loss of biological activity. The collaborative study clearly illustrated how sparse existing data were concerning the performance of IL-8 in different biological assays. Primarily, the study showed that the most popular assays for IL-8, namely chemotaxis of neutrophils, were so variable that no consistent or statistically valid estimates of potency could be deduced from them. In addition, unexplained differences in the relative activities of certain preparations were seen. The relative activity of two preparations depended on the assay system (Fig. 2). Considering that all the preparations included in the study were reported by the manufacturers to be identical, this suggests that an unknown structural component must play a role in IL-8 biological responses.

The inclusion in the design of the study of reporting whether material was used fresh or frozen illustrated that (i) IL-8 is a molecule which is highly unstable to freeze-thawing and (ii) the extent to which this occurs is preparation-dependent. Therefore, several previously unknown scientific findings produced from the collaborative study provided criteria for the choice of material to serve as an international standard. The selected preparation did not exhibit variable relative potency in different assays and was stable even after freeze thawing. In fact, the problem of stability was so great that it has been recommended in the memorandum sheet accompanying the standard that it be stored at 4°C for only four weeks once reconstituted.

Case Study 3: The International Collaborative Study of Interferon alpha

IFN-alpha is a family of potentially 28 closely related proteins that vary from each other by a few amino acids. IFN-alpha has potent anti viral and antiproliferative activities and has been commercially available for many years to treat viral-based diseases. In addition, it is used as an immunomodulator to treat diseases such as cancer [6]. Several WHO international standards for IFN-alpha were established in the 1980s, including a «primary» international reference preparation of leucocyte interferon and several subtype specific international standards. Subsequently, problems arose in the calibration of IFN-alpha preparations that appeared to originate from discontinuity between the various international standards. To resolve the situation, several unanswered scientific questions needed to be addressed [10].

One vital premise for standardization is to compare like with like. However, it was not known which, if any, IFN-alpha subtypes could be compared in this way and what standard preparation should be used. In addition, there had been indications that different antiviral assay cell/virus combinations detected differences in the biological activities between IFN-alpha preparations. Therefore, it had to be

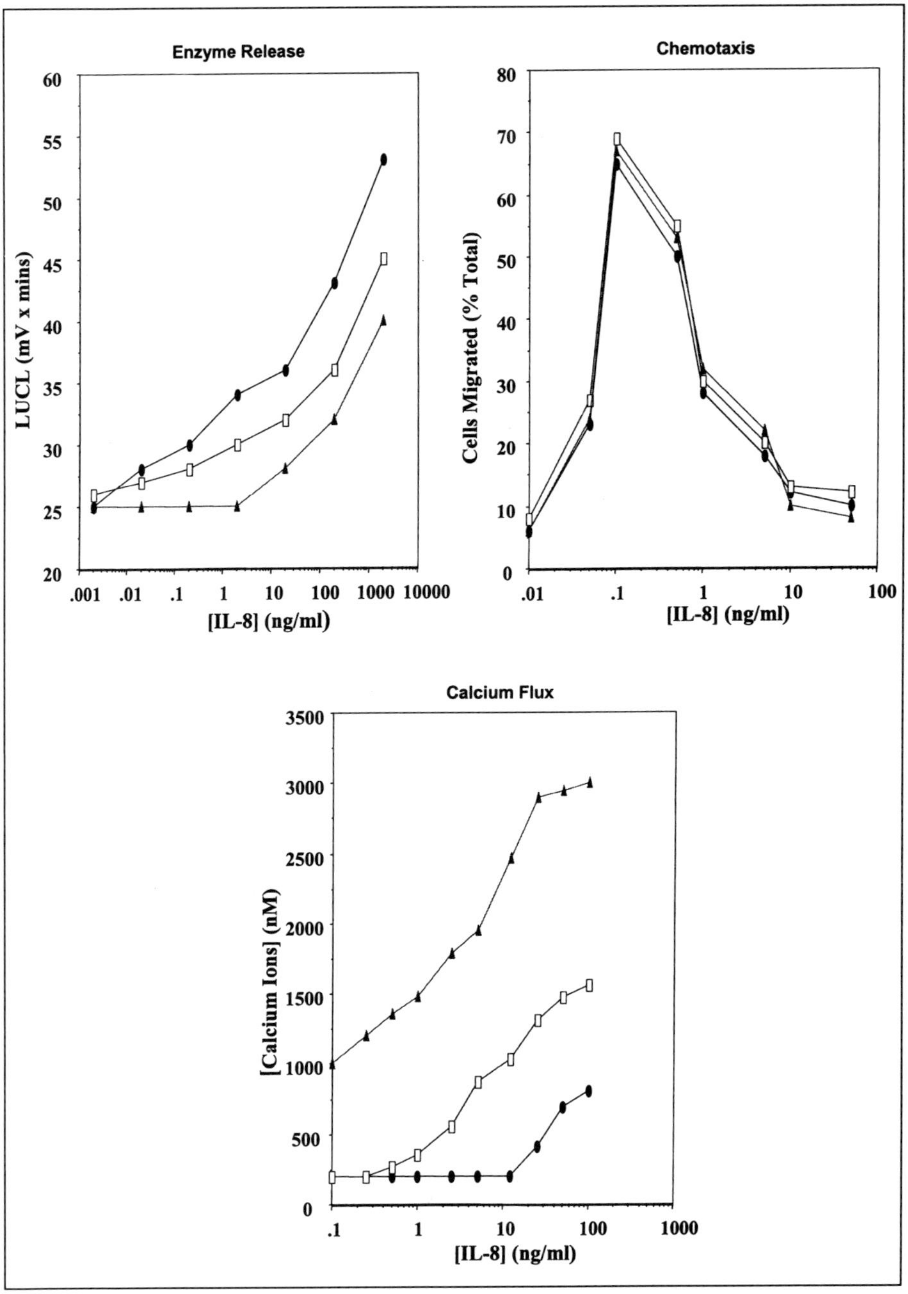

Fig. 2: Results from the International Collaborative Study for Interleukin-8. Three different preparations of IL-8 were bioassayed in the three formats of neutrophil based bioassay illustrated. This illustrates the variable relative potencies of preparations labelled ▲ and ● to preparation ❑, dependent on the biological activity being measured.

established whether the use of different bioassay formats results in different relative activities of the various IFN preparations available.

The most popular biological assays for interferons were based on their antiviral activity, but the advent of therapeutic applications using the immunomodulatory activity of interferon alpha has led to an increasing development and use of antiproliferative or immunomodulatory bioassays. However, it was not known before the collaborative study whether there was a consistent ratio of antiviral to antiproliferative activity of the interferon alpha subtypes and thus whether the results of the two groups of bioassays were comparable.

The results of the collaborative study clearly showed that the relative activities of the interferon alpha preparations depend on the assay systems in which they are compared and on the similarity of the preparations. The study illustrated that the greater the molecular dissimilarity between two interferon alpha preparations, the greater the differences in relative activity in different assay systems. For example, the comparison of a recombinant preparation of interferon alpha 2b to the current International Reference Preparation of natural leukocyte interferon, 69/19 was highly variable depending on the antiviral cell/virus combination used (Fig. 3A). However, comparisons of a leukocyte preparation with the leucocyte IRP 69/19 were unaffected by the different assays, highlighting that the like with like approach results in standards that can be used independently of the type of bioassay used (Fig. 3B). It was also shown that the interferon alpha subtypes exhibit different ratios of antiviral to antiproliferative activity, again illustrating that the type of bioassay used to measure the potency of interferon alpha influences the results produced. The design of the collaborative study without inclusion of the scientific rationale to answer the questions about the basic biology of interferon alpha would have led to a continuation and probable expansion of this highly confused area of standardization and the possible establishment of inappropriate standards. However, once the basic relationship of biological activity to protein structure and assay type was confirmed by the study, there were clear indications of the standards required and how they should be used.

Conclusions

A WHO international cytokine standard serves as the primary reference preparation in terms of which all other reference standards should be calibrated. It is necessary to ensure that the material contained in the international standard has been shown to be suitable to serve as a bioassay standard. It is therefore vital to carry out international collaborative studies on a range of materials from different sources to be tested in a variety of assays, by a representative spread of laboratories worldwide. If a suitable standard is to be established, the design of such studies must take into account scientific issues surrounding cytokine biological activity on a case-by-case basis dependent upon the cytokine in question.

Acknowledgements

We would like to thank all the manufacturers and research institutions who have generously provided cytokines to the WHO programme. In addition we gratefully

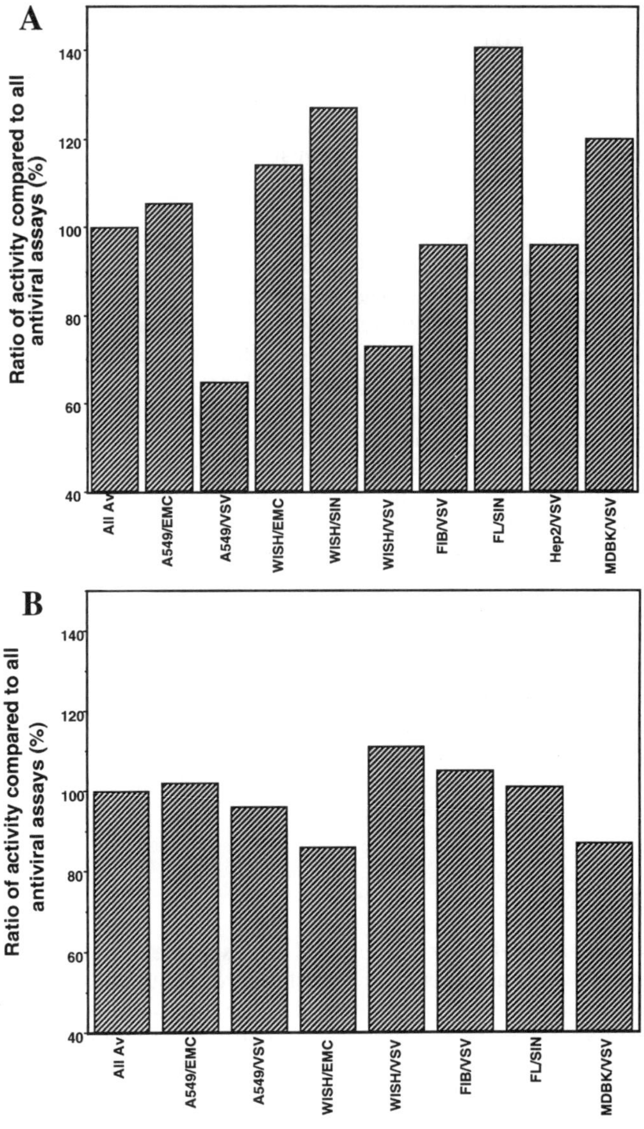

Fig. 3: Results form the International Collaborative Study for Interferon alpha.
Panel A. A preparation of recombinant Interferon alpha 2b was assayed against the International Reference Preparation of natural leukocyte interferon, 69/19, in the various antiviral assay cell/virus combinations illustrated. This clearly shows that different cell line/virus assays can distinguish different relative activities between these two dissimilar preparations.
Panel B. A preparation of natural leucocyte interferon alpha was assayed against the International Reference Preparation of natural leukocyte interferon, 69/19, in the various antiviral assay cell/virus combinations illustrated. This clearly shows that different cell line/virus assays cannot distinguish different relative activities between these two similar preparations.

acknowledge the contributions of the participants in the various WHO international collaborative studies and the Standards Division at NIBSC for their dedication to the production of cytokine standards.

References

1 Mire-Sluis AR, Thorpe R (eds): Cytokines. London, Academic Press, 1998.

2 Thomson A (ed): The Cytokine Handbook. London, Academic Press 1998.

3 Oppenheim JJ, Rossio JL, Gearing AJH (eds): Clinical applications of cytokines: Role in pathogenesis, Diagnosis and Therapy. Oxford, Oxford University Press, 1993.

4 Steward WP: Granulocyte and granulocyte colony stimulating factor. Lancet 1993;342:153-157.

5 Erslev AJ, Adamson JW, Eschbach JW, Winearls C (eds): Erythropoietin: molecular, cellular and clinical biology. London, Johns Hopkins University Press, 1991.

6 Meager A: Interferons alpha, beta and omega, in Mire-Sluis AR, Thorpe R (eds): Cytokines. London, Academic Press, 1998, pp 361-389.

7 Mire-Sluis AR, Gaines Das R, Padilla A: Facilitating the development of cytokines in research and as therapeutic agents: The WHO Cytokine Standardisation Programme. J Imm Methods 1998;216:103-116.

8 Mire-Sluis AR: Expression of potency: Why units not mass? Pharmaceutical Sciences 1997; 3:15-18.

9 Mire-Sluis AR, Gaines-Das R, Gerrard T, Padilla A, Thorpe R: Biological Assays: Their role in the development and quality control of biological medicinal products. Biologicals 1996;24:351-361.

10 Mire-Sluis AR, Gaines Das R, Zoon K, Padilla A, Thorpe R, Meager A: The biological properties, assay and standardisation of Interferon-alpha: A need for a WHO Collaborative Study. J Interferon and Cytokine Research 1996;16:637-643.

11 Mire-Sluis AR, Gaines Das R, Thorpe R: The international standard for granulocyte colony stimulating factor (G-CSF): Evaluation in an international collaborative study. J Immunol Methods 1995;179:117-126.

12 Mire-Sluis AR, Gaines Das R, Thorpe R: Implications for the assay and biological properties of interleukin-8: Results of a WHO International Collaborative Study. J Immunol Methods 1997; 200:1-16.

Dr. A.R. Mire-Sluis, Division of Immunobiology, National Institute for Biological Standards and Control, Blanche Lane, South Mimms, Potters Bar, Herts. EN6 3QG, U.K.
Now at: Genentech, Med-Bioanalytical Methods Development, 1 DNA Way, Mail Stop 38, South San Francisco, CA 94080, USA.

Brown F, Griffiths E, Horaud F, Schild GC (eds): A Celebration of 50 Years of Progress in Biological Standardization and Control at WHO.
Dev Biol Stand. Basel, Karger, 1999, vol 100, pp 95-99

Challenges for Biological Standardization and Control in the 21st Century

K.C. Zoon

Center for Biologics Evaluation and Research, Bethesda, MD, USA

> «Science and Peace will triumph over ignorance and war, nations will unite, not to destroy, but to build, and the future will belong to those who have done the most for suffering humanity.»
>
> *Louis Pasteur*

This quote in the foyer of the Center for Biologics Research and Review (CBER) has been an inspiration to many scientists in our organization who have worked diligently with the World Health Organization (WHO) and other institutions all over the world to better the human condition. Biological standardization and control have been and will continue to be important elements in achieving this goal.

The regulation of biological products is founded in science and law. It encompasses many activities, including review, research, policy, compliance and product surveillance (Fig. 1). Each is equally important to ensure that safe and effective biological products are available. Underpinning our efforts is strong and critical scientific thinking and decision-making. Regulatory research is one critical element. The functions of regulatory research are shown in Table 1. However, each element is important for ensuring biological product quality, safety and efficacy. Review activities are essential to appropriate regulation of biological products. These include the review of products during development, e.g. investigational new drug and device applications, master files, and finally licensing and premarket applications. Postmarketing surveillance is an important sentinel system to ensure the safety of the product. Careful monitoring and appropriate actions are necessary to ensure that patients and physicians are aware of the side effects of the product or that the product is removed from the market if warranted. Compliance includes education, assessment of a manufacturer's product through inspections, product testing, protocol review and enforcement actions. The development of policy is also critical for promoting and enabling biological product advancement and providing a clear understanding of the law and the expectations of a regulatory agency.

The biological products regulated by CBER are shown in Figure 2. Each major area poses unique challenges and provides important public health benefit. Our traditional areas of regulation include blood, blood components, blood diagnostics,

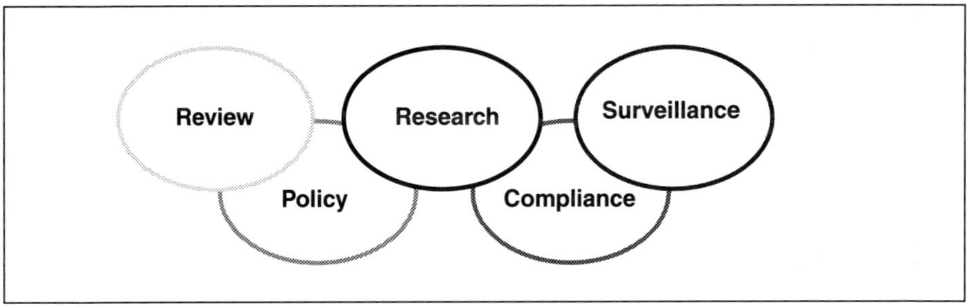

Fig. 1: Regulation of biological products.

Table 1: Functions of regulatory research.

• Facilitate the approval of safe and effective products
• Support decisions to withdraw products that are found to be unsafe
• Anticipate public needs and support informed decision-making in prevention of, and response to, public health crises
• Encourage industry-wide adoption of new technologies
• Facilitate development of industry-wide standards and methods
• Contribute to improvement of existing products and the development of new products
• Aid in recruitment and retention of excellent scientists

blood derivatives, vaccines, venoms and allergenic products. Our «conventional» biotechnology products include monoclonal antibodies and recombinant DNA derived proteins. Newer areas of biological products encompass somatic cell and gene therapy, tissues, and xenotransplantation. Areas needing consideration in the future include cloning technology, germ-line gene therapy, or in-utero gene therapy. Each of these poses not only scientific challenges but also social, ethical and legal challenges. A number of cross-cutting considerations exist for all biological products; they include concerns regarding contamination with adventitious agents, the quality of the source material, and the need to fully control and validate the manufacturing process.

There are a number of major forces shaping biological medicines in the 21st century (Table 2). Large investments by government, industry and private institutions in new biological product discovery is already yielding many important new products and will continue to do so into the next century. This is truly the age of biology, from the new therapeutics and diagnostics derived from the human genome project to the development of new vaccines, e.g. DNA vaccines and plant-derived vaccines, to improvements in blood safety, e.g. gene amplification testing for viral contamination.

A number of technological developments have facilitated the characterization of biological products. These include in vitro, biochemical and analytical assays. An example is the mutant analysis by PCR and restriction enzyme cleavage

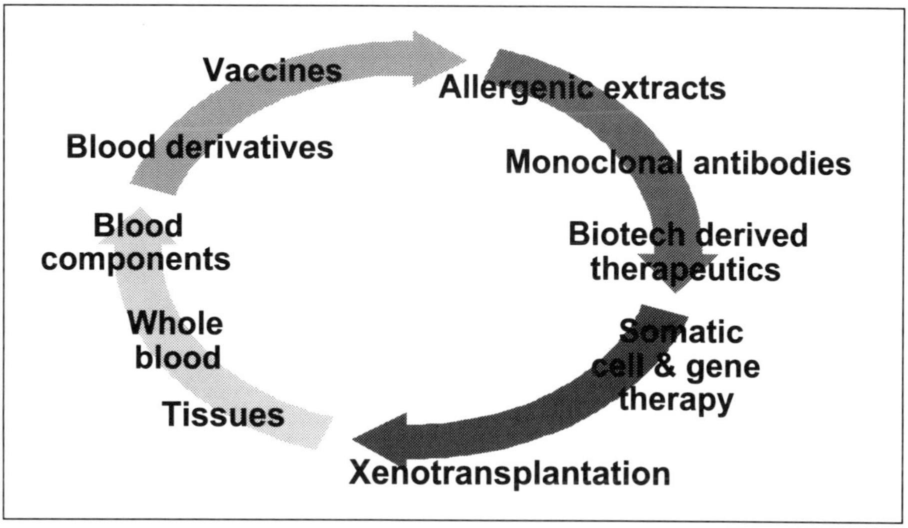

Fig. 2: Biological products regulated by CBER.

Table 2: Forces shaping biological medicines in the 21st century.

• New discovery research and technology
• Immunization for disease prevention
• Demand for new medicines
• Faster access
• Global market: International harmonization and competition
• Changing health care environment

(MAPREC) assay. It has been developed at CBER to detect a site for type 3 poliovirus that is critical in determining the neurovirulence of the virus, i.e. nucleotide 472. If the level of change of U to C exceeds 1%, then the virus will be positive in the monkey neurovirulence test. The test is currently being incorporated into the WHO requirements for live attenuated Oral Polio Vaccine to help assess the safety and consistency of the vaccine. New analytical tools include matrix assisted laser desorption ionization time of flight (MALDI-TOF) spectroscopy, capillary electrochromatography, signal transduction fingerprinting and microarray technology. Microarray technology (e.g. DNA chip) is an important new tool to monitor gene and protein expression.

The need for global immunization to prevent major infectious diseases is another driving force shaping the biologicals field. From the global eradication of poliomyelitis to the challenges of developing new safe, effective and accessible vaccines for tuberculosis, HIV and malaria, vaccines are key biological products that are needed to improve public health.

Demands for safer blood products will continue to grow into the next century. New measures to: (i) screen blood donors; (ii) test blood and blood components, and derivatives, and (iii) inactivate/remove adventitious agents from these products are all rapidly advancing and should help facilitate even safer whole blood and blood products in the future.

The need for safe and effective new biological products made available in a timely way will also be a major force shaping biological medicines in the next century. Control authorities, industry, academia, and health care providers will be under increased pressure by the public to quickly translate new biomedical discoveries into widely available medicines. For the regulatory authorities it will result in a demand for guidance and standards to enhance clinical development and premarketing review performance that leads to the timely availability of these products.

The biological product development process is shown in Figure 3. The Food and Drug Administration Modernization Act of 1997 (the Act) has several directives that are intended to: (i) shorten drug development time, i.e. manufacturing, preclinical and clinical studies; (ii) shorten license/approval times; and (iii) increase access to new medicines and information about them. An important component of the Act is the establishment of a new programme called Fast-Track. This programme is intended to decrease the development time of new medicines for severe and life threatening conditions for which there are unmet medical needs. The programme fosters close interactions of the sponsor with the FDA and performance enhancing measures. A key challenge while implementing these performance enhancements is to ensure a strong post-marketing surveillance programme; this includes monitoring of adverse events, lot release testing, education, inspections and enforcement.

Quality and safety issues associated with biological products are very important. Characterization of the source material, adequate production processes and control

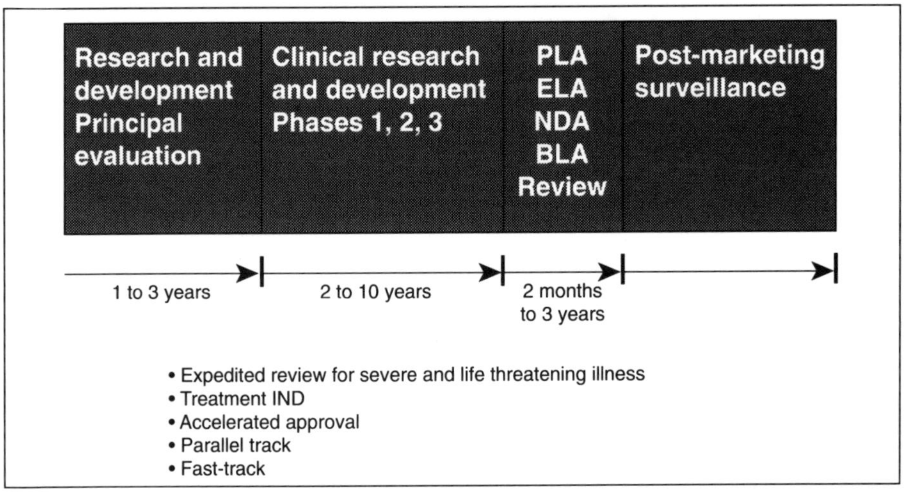

Fig. 3: New biological drug development.

testing, and process validation are fundamental to ensuring the quality and consistency of biological products. The need to adequately detect, remove/inactivate adventitious agents (such as bacteria, viruses, mycoplasma, fungi, parasites, and new infectious agents not yet completely identified, i.e. TSE) will continue to be a challenge. Application of new technologies to the detection of adventitious agents, e.g. polymerase chain reaction (PCR) and related assays provides the opportunity to enhance biological product safety. Standards for licensing biological products will continue to ensure that they are safe, pure, potent, effective and are produced under conditions which meet current good manufacturing practices.

As we approach the next century, the number of new biological products under clinical investigation are increasing (Fig. 4). Over half of the new biological products being studied are derived from biotechnology. New review performance processes for license applications have reduced the time to license new biological medicines from an average of 24 months to 12 months.

In conclusion, the challenges for the 21st century for biological products are many. They include new technology/discovery, safety of biological products, ethical issues e.g. cloning, germ-line gene therapy, access to biological products, especially in developing countries, and harmonization of regulatory standards. These challenges will require increased international cooperation. The World Health Organization will play a vital part in the resolution of many of these issues. We at CBER look forward to working with WHO to meet these challenges.

Dr. K. Zoon, Center for Biologics Evaluation and Research, Rockville Pike, Suite 200N, Bethesda, MD 20852, USA.

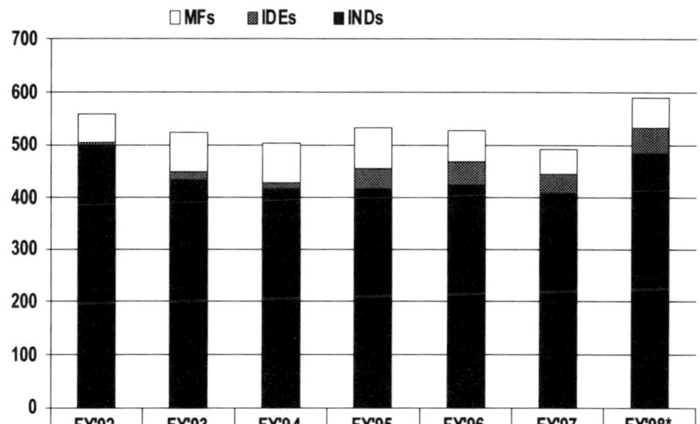

	FY'92	FY'93	FY'94	FY'95	FY'96	FY'97	FY'98*
MFs	53	74	75	79	60	47	60
IDEs	9	17	12	38	43	36	48
INDs	496	432	416	414	424	406	482

* FY98 Projected: 1 Oct 97-31Aug 98 (INDs = 442, IDEs = 44 and MFs = 55).

RIMS:9/21/98

Fig. 4: CBER INDs/IDEs/MFs received FY'92-FY'98*.

Brown F, Griffiths E, Horaud F, Schild GC (eds): A Celebration of 50 Years of Progress in Biological Standardization and Control at WHO.
Dev Biol Stand. Basel, Karger, 1999, vol 100, p 101

SESSION III

Future Challenges

Co-chairs: *Kathryn Zoon (USA)*
and Victor Grachev (Russia)

Brown F, Griffiths E, Horaud F, Schild GC (eds): A Celebration of 50 Years of Progress in Biological Standardization and Control at WHO.
Dev Biol Stand. Basel, Karger, 1999, vol 100, pp 103-118

The Transmissible Spongiform Encephalopathy Agents: Concerns and Responses of United States Regulatory Agencies in Maintaining the Safety of Biologics

D.M. Asher

Laboratory of Method Development, Division of Viral Products Office of Vaccine Research and Review, CBER/USFDA, Rockville, MD, USA

Abstract: Regulatory agencies responsible for protecting public health must be concerned with reducing or preventing opportunities for exposure of humans and animals to the agents of transmissible spongiform encephalopathies (TSEs), especially bovine spongiform encephalopathy and Creutzfeldt-Jakob disease. The United States Food and Drug Administration (FDA) has taken precautionary actions to assure that regulated products are free of such infectious agents, from both animal and human sources, including the issuing of a regulation and a number of guidance documents. With regard to TSEs, the materials of greatest concern to FDA's Center for Biologics Evaluation and Research have been bovine gelatin, tallow derivatives and serum, as well as human-derived products and excipients including blood, blood components and plasma derivatives. A number of newer issues regarding TSEs now confront agencies responsible for protecting public health. It would be of great help to have standard reference materials available to assist in the diagnosis of the diseases and to detect the infectious agents.

Introduction

This paper summarizes recent concerns and actions of the United States Food and Drug Administration (FDA) regarding the transmissible spongiform encephalopathies (TSEs or prion diseases [1]) and the safety of products regulated as biologicals in the U.S.A. as of 27 October 1998. (It attempts to explain FDA policies but should not be considered as official FDA Guidance [2]). In addition, it briefly addresses several issues posed by the TSEs that may soon merit attention by national regulatory authorities and international agencies dedicated to improvement of human and animal health.

Although very rare, the TSEs have occasioned great public anxiety because they are uniformly fatal diseases, most frequently accompanied by dementia. The fact that they result from «slow» infections that may lie dormant for years after infection was acquired and are caused by a unique type of pathogen, the nature of which remains shrouded in mystery, increases the sense of dread that they inspire.

Concerns of the FDA regarding the TSEs fall into two general categories (Table 1): (i) TSEs of animals potentially transmissible to other animals or to humans via FDA-regulated products; (ii) TSEs of humans either known to have been transmitted from one person to another (causing iatrogenic Creutzfeldt-Jakob disease) or to be potentially transmissible in that way. FDA (and the United States Department of Agriculture [USDA]) regulations and guidance policies regarding TSEs may be viewed as attempts to reduce opportunities for human and animal exposure to products through control of critical points: sourcing of raw materials, processing during manufacture and end use of products.

Recently, the animal TSE of greatest concern to the FDA and to other national regulatory authorities has been bovine spongiform encephalopathy (BSE), because BSE is the probable source of new-variant CJD (vCJD or nvCJD) in humans [3]. The most important actions taken by agencies of the U.S. government to reduce the risk of BSE for animals and humans have been in the form of regulations: (i) USDA restricted the importation of cattle and many bovine products from an increasing number of European countries to reduce opportunities for introducing the BSE agent into U.S. cattle [4]; (ii) To prevent the feed-borne spread of BSE, should the agent accidentally infect U.S. ruminants, the FDA (through the Center for Veterinary Medicine) prohibited the feeding of most mammalian proteins to ruminants and has required labelling of feeds containing such proteins intended for other animals to indicate that they contain materials that must not be fed to ruminants [5].

This article concentrates on FDA policies regarding TSEs and the safety of biologicals, with some attention to other pharmaceuticals and to devices, because those are the products that most concern the FDA Center for Biologics Evaluation and Research (CBER), the WHO Biologics Programme, the International Association of Biological Standardization, and this journal. The policies described were all promulgated in the form of guidance representing the current opinion of the FDA; such guidance is not, in itself, legally binding on either the FDA or industry [2]. The

Table 1: TSEs and biologicals: FDA concerns and actions.

FDA goal
• Prevent transmission of disease via regulated products.
Recent FDA concerns
• BSE and new-variant CJD • Other forms of CJD (iatrogenic transmission)
Recent FDA actions
• Ruminant «feed ban» (regulation) • Bovine gelatin (guidance) • Bovine tallow, tallow derivatives (policy review) • Bovine serum (guidance) • «CJD-implicated» human blood, blood components and plasma derivatives (guidance)

FDA policies reviewed here are those addressing the safety of bovine gelatin, bovine tallow derivatives and bovine serum as well as human blood and its components and human plasma derivatives, because those materials are either components in CBER-regulated products or, in the case of plasma derivatives, are themselves primary biological products.

General Review

In the U.S.A., all recognized accidental transmissions of CJD to humans [6] have been attributed to exposure to products of human origin (Table 2), products that happen to belong to classes of products regulated by the FDA (dura mater allografts [7-10], corneal transplants [11] and human cadaveric pituitary hormones [12, 13]). A number of iatrogenic transmissions have also occurred from exposure to contaminated neurosurgical instruments [14-16], some of which are FDA-regulated by the Center for Devices and Radiological Health.

Although relatively few iatrogenic transmissions of human TSEs have been recognized in the U.S.A., the potential for spreading such infections via regulated products weighs heavily on the minds of authorities. The largest recognized outbreak of a human TSE, kuru, which affected hundreds of women and children as well as a smaller number of men through the 1980s, clearly resulted from exposure to infected human tissues; transmission of the disease was eliminated by ending the exposure [17-19].

No vaccine or other product regulated as a biological in the U.S.A. (except for cornea, regulated as a device until recently reclassified) has been implicated in an accidental transmission of CJD. (Pituitary hormones are peptides regulated as drugs and dural allografts are currently regulated as devices.) But it is cautionary to recall the role that a veterinary vaccine played in the early history of scrapie research. In early 1935, W.S. Gordon at the Compton Field Station in England had completed successful trials with a new vaccine against louping ill – a common tick-borne infection of sheep in the UK caused by a small positive-sense RNA flavivirus [20]. The vaccine was a formalin-inactivated suspension of brains, spinal cords and spleens from sheep experimentally infected with the virus. All pilot lots of vaccine had passed conventional safety tests, causing no acute illness in mice or sheep inoculated intracerebrally; and, in field trials, more than 1000 sheep had already been effectively protected against louping ill without any problems. In March, Gordon distributed three large bulk batches of vaccine – some 45,000 doses in all. Batches 1

Table 2: Iatrogenic transmission of CJD by regulated products of human origin.

Product	Number of patients	Mean incubation	Range of incubations
Cornea	2	17 mo	16 and 18 mo
Dura mater graft	47	7.4 yr	1.3 to 16 yr
Pituitary extracts • growth hormone • gonadotrophin	 > 95* 4	 12 yr 13 yr	 5 to 30 yr 12 to 16 yr

* 16 cases/~8000 estimated recipients in U.S.A.

and 3 performed as well as the pilot lots. Two and a half years later, surviving sheep immunized with vaccine from batch 2 began to show signs of a progressive neurological disease – not the encephalomyelitis typical of louping ill but the TSE, scrapie. Some affected flocks had never before had a recognized case of scrapie; now the incidence reached as high as 35%. Gordon traced back 112 source sheep used to prepare the bad batch of vaccine; eight were progeny of healthy ewes that had been in contact with scrapie animals. Several of those ewes eventually got scrapie themselves [21].

That episode gave the first indication that the scrapie agent was unusually resistant to chemical inactivation. It confirmed earlier experimental studies showing that scrapie had very long asymptomatic incubation periods [22] during which tissues of healthy looking animals are infectious. It should have served as a warning of what can happen if a TSE agent contaminates the source material used to prepare a pharmaceutical. Somehow, that lesson was ignored, and the same kind of accident happened again – several times – with human cadaveric pituitary hormones [6, 12, 13].

TSEs and Products of Animal Origin

In the U.S.A., no human TSE has been convincingly attributed to exposure to an animal TSE agent; surveillance by USDA has not detected BSE in U.S. cattle [23], and the Centers for Disease Control and Prevention (CDC) has not confirmed any case of nvCJD in anyone. [24, 25]. Other countries have been less fortunate. The BSE outbreak peaked in the UK at the end of 1992 [26], and worldwide anxiety was abating until March 1996, when the first report appeared of nvCJD (Table 3 [3, 27-29]), a disease most plausibly attributable to human exposure to the BSE agent. (In retrospect, recognition of a new spongiform encephalopathy in domestic cats and zoo felines in the UK [30], might have served to alert authorities that the agent of BSE had a broader host range than the agent of sheep scrapie – an agent never implicated in disease of cats or humans – and that the probable vehicle of transmission of BSE to cats was meat or a meat product.) The attribution of nvCJD to infection with the BSE agent greatly increased concern within the FDA and accelerated regulatory efforts. Within weeks, a new proposed rule, intended to prevent the spread of BSE in cattle by prohibiting use of most mammalian proteins (usually meat-and-bone meal) in ruminant feeds, was announced [31]; a final rule – «feed ban» – was issued in just over one year [5].

USDA was first to articulate concern about dangers of transmitting BSE to U.S. ruminants via imported gelatin, noting, in December 1991, that bovine gelatin from BSE countries is «...not to come in contact with ruminants. (and importers of gelatin from BSE countries must obtain veterinary permits). for Importation and Transportation of Controlled Materials and Organisms and Vectors... [32]». In September 1993, USDA further explained that «...[G]elatin derived from ruminants (from BSE countries) poses a risk of spreading BSE (to ruminants) [33].»

Even before the early reports of nvCJD appeared, FDA became concerned about potential transmission of the BSE agent to humans, noting, in letters to manufacturers of dietary supplements in November 1992 and to manufacturers of drugs, biologicals and medical devices in December 1993, that «FDA believes that BSE is a concern. ... [that] precautionary measures will reduce potential risk of

Table 3: Clinical and histopathological features of patients with new-variant CJD and typical sporadic CJD[a].

Feature	New-variant CJD	Sporadic CJD
Age[b] at death in yr average (range)	29 (19-41)	65
Duration[b] of illness, mo average (range)	12 (8-23)	4
Presenting signs	Abnormal behavior Dysesthesia	Dementia
Later signs	Dementia Ataxia Myoclonus	Ataxia Myoclonus
Periodic complexes on EEG	None	Most
PRNP Codon 129 Met/Met	100%	83%
Histopathological changes	Vacuolation Neuronal loss Astrocytosis Amyloid plaques (100%) «Florid» PrP plaques[c] 100%	Vacuolation Neuronal loss Astrocytosis Amyloid plaques (~15%) No «florid» PrP plaques[c]
Mice inoculated with brain suspensions	BSE-like lesion profiles[d]	Non-BSE-like lesion profiles
PrP glycosylation pattern	BSE-like[e]	Not BSE-like

a 10 patients with v-CJD and 185 with sporadic CJD [3].

b Median age and duration for v-CJD, average for typical CJD.

c Dense plaques, pale periphery, halo of spongiform vacuoles.

d [84].

e «Type 4» PrP [85]: high molecular-mass PrP glycoform (diglycosylated) exceeds mono-glycosylated and non-glycosylated forms in relative abundance.

exposure, ... [and] that it is prudent to reduce any potential risk of human exposure to the BSE agent. ... [The FDA] request[ed] that bovine-derived materials from cattle that have resided in or originated from USDA-designated in or originated from USDA-designated BSE countries not be used in the manufacture of FDA-regulated products intended for humans or animals [34].» However, in letters to trade organizations of gelatin manufacturers in July 1994, FDA clarified that it did «...not object to the use of bovine-derived materials from BSE-countries in the manufacture of pharmaceutical grade gelatin, ... [although] ... [the FDA] continue[s] to consider it prudent. ... to obtain such materials from non BSE-countries whenever practical [35].» In a published letter to manufacturers and importers of dietary supplements and cosmetics, FDA elaborated that «At this time, FDA is not extending the recommendation ... to dairy products or gelatin, because available evidence does not suggest transmission via these foods [34].» The exemption for gelatin from BSE-free sourcing recommendations reflected a conclusion by FDA that conditions for manufacturing gelatin were likely to inactivate the infectious agent [36]. Regulators may also have relied implicitly on a widely held view that a «species barrier» (the relative resistance of animals of one species to infection by a TSE agent adapted to another species of animal not closely related) between cows and humans would protect humans from infection with the bovine-adapted BSE agent, as such a species barrier was thought to have protected humans against infection with the sheep-adapted scrapie agent.

In 1996, FDA reconsidered the safety of gelatin imported from BSE Countries, for three reasons: (i) New-variant Creutzfeldt-Jakob Disease was recognized in persons in the UK and France (reducing confidence that a species barrier reliably protected humans from infection with the agent causing BSE); (ii) Experimental data submitted to the agency failed to show that methods used for gelatin processing had removed all TSE infectivity present in starting materials [37]; (iii) FDA became concerned that some source materials for imported gelatin might contain neural tissues of cattle from BSE countries [37].

To address the issue of safety of gelatin and gelatin by-products derived from sources potentially contaminated with BSE agent, FDA convened a meeting of its TSE Advisory Committee (TSEAC) in April 1997. The TSEAC reached eight conclusions: (i) current scientific evidence no longer justifies exempting gelatin from restrictions recommended by FDA for other bovine-derived materials originating from BSE countries; (ii) the USDA «BSE List» should be expanded to identify countries that, although not reporting BSE in native cattle, have surveillance systems inadequate to assure that BSE is not present; countries of unknown status should be considered less reliable sources of bovine-derived materials than «BSE-free» countries with adequate surveillance systems; (iii) bovine gelatin administered parenterally poses a greater risk of transmitting TSE to humans than would the same product when ingested [38]; (iv) brains and spinal cords of cattle from BSE countries should be excluded from raw materials used to produce gelatin for human consumption; (v) Alkaline processing with lime may reduce amounts of infectious TSE agent in gelatin substantially but has not been demonstrated to eliminate infectivity completely; acid processing alone appears to be much less effective; (vi) Other steps used in the manufacture of gelatin – degreasing, neutralization with caustic soda, filtration, deionization and heat sterilization – might also reduce levels of infectivity but cannot be relied upon until validated; (vii) Expert consultation should be sought when designing process-validation studies for gelatin; (viii) Porcine gelatin poses no known risk of transmitting TSE to humans.

Based on the data presented to TSEAC as well as discussions and opinions expressed by members of the committee, FDA issued new Guidance for Industry on 7 October 1997 [35]. It consists of seven points, summarized here: (i) Determine tissue, species and country source of gelatin raw materials; (ii) Bones and hides of cattle showing signs of neurological disease should not be used to manufacture gelatin; (iii) Gelatin from bones and hides of cattle from BSE countries or countries of unknown BSE status (by standards of the Office International des Epizooties [39]) should not be used in injectable, implantable or ophthalmic products regulated by the FDA; (iv) «At this time» FDA does not object to oral and cosmetic use of gelatin from bones of cattle from BSE countries if cattle were «... from BSE-free herds and if. ... heads, spines and spinal cords [were removed] directly after slaughter.»; (v) FDA does not object to bovine hide gelatin for foods and cosmetics if hides of cattle with signs of CNS disease were excluded and contamination of hides with CNS and eye tissues was avoided; (vi) FDA does not object «... to use of any bovine gelatin from U.S. animals or animals from other BSE-free countries…»; (vii) FDA does not object «…to use of pig-skin gelatin. .. uncontaminated with bovine materials from BSE countries or countries of unknown BSE status [by OIE standards].»

The TSEAC was reconvened in April 1998 to review, among other issues, FDA's guidance policy concerning imported bovine gelatin for oral and topical use. Most TSEAC members generally agreed with the FDA guidance on gelatin, although a majority believed that it might still be safe to remove the spinal columns from carcases some time later rather than immediately after slaughter. FDA is currently reviewing policies concerning safe sourcing of raw materials to manufacture gelatin.

In April 1998, the TSEAC also reviewed FDA Policies on safety of bovine tallow and tallow derivatives in regulated products, which are, in accordance with the letters of 1993 [34], recommended to be prepared from raw materials sourced exclusively from animals in BSE-free countries. The TSEAC was reassured that processing conditions described by industry representatives – who reported that high temperatures, elevated pressures and prolonged times of exposure to heat and pressure were universally applied during manufacture of tallow derivatives [40] – were sufficiently rigorous to eliminate all detectable infectivity of TSE agents and that the effectiveness of those methods had been sufficiently validated [41] so that tallow derivatives from any country might safely be used in FDA-regulated products; however, a majority of the TSEAC members expressed the opinion that beef tallow itself should not be sourced from tissues of animals from BSE countries. FDA is considering revised policies concerning the safe sourcing of bovine tallow derivatives.

Concerning safety of bovine serum from sources potentially contaminated with BSE agent [42], FDA regulations (21 CFR 618 et seq) require that «Cultures used in the manufacture of products … shall be … free … from extraneous organisms … [and] … tested for the presence of detectable microbial agents … [as] … necessary to assure the safety, purity, and potency of a product…». In July 1993, CBER issued revised Guidance on Characterization of Cell Lines Used to Produce Biologicals (Memorandum to Manufacturers, Docket 84N-0154) recommending that «… serum or additives … [in] … culture … medium should be certified to be free from contaminants and adventitious agents, such as the agent responsible for the production of Bovine Spongiform Encephalopathy.» Furthermore, the issue of using bovine serum

Table 4: Comparison of current FDA policies on safe sourcing of three regulated products from BSE-free and BSE countries (includes BSE-status-unknown countries).

Route	Gelatin	Tallow	Tallow derivatives
Injectable etc.[a]	BSE-free[b]	Not used	BSE-free[c]
Oral (food, drug)	BSE+precautions[d]	BSE-free	BSE-free[c]
Topical	BSE+precautions[d]	BSE-free	BSE-free[c]

[a] Etc. = implantable and ophthalmic routes

[b] TSEAC advice on this route was not solicited by FDA.

[c] TSEAC advised that source from a BSE country was acceptable if the manufacturing process was adequate (at least one exposure to ≥ 133 C for ≥ 20 min at ≥ 3 bar pressure) and validated.

[d] TSEAC advised one slightly relaxed precaution noted in text.

from BSE countries in FDA-regulated products in the U.S.A. has effectively been rendered moot by regulations of the USDA interpreted by the Animal and Plant Health Inspection Service (APHIS) to prohibit importation into the U.S.A. of bovine serum from countries with BSE for purposes other than scientific, educational and research, for which a permit is required [9 CFR 95.4]. To the knowledge of CBER, no regulated biologic is currently prepared using bovine serum sourced from animals in a BSE country. Furthermore, on 12 December 1997, pending clarification of the status of European countries, as a preventive step, the USDA prohibited importation of all live ruminants and most ruminant products (excluding gelatin for human consumption, milk and milk products) from all countries of Europe due to potential risk of BSE [4]. Therefore, at the time of this writing, APHIS policy effectively prohibits the legal importation into the U.S.A. of bovine serum from all European countries, pending receipt by the USDA of information showing a country to be at low risk for BSE in its cattle.

Actual experimental inoculations of susceptible animals with blood from bovines with BSE have not detected infectivity [43]; similar studies with blood of sheep with scrapie yielded similar results [44]. However, the FDA has continued to consider it prudent to avoid using bovine serum from BSE countries to manufacture regulated products, including pediatric vaccines. FDA's very conservative policy has apparently not interfered with the supply of such vaccines.

TSEs and Biologicals of Human Origin (Blood, Blood Components and Plasma Derivatives)

There is conflicting evidence to suggest that blood of humans and animals may sometimes contain infectious TSE agents. As noted above, most studies of naturally infected animals and humans with TSEs failed to detect the transmissible agent in blood; however, in some patients dying with CJD, infectivity has been found in spleen, lymph nodes and liver [45], raising the concern that infected cells from those tissues may enter the circulation. Furthermore, in several species of rodent experimentally infected with scrapie or CJD agents, small amounts of infectivity were detected in blood [46-48] (≤ 10 rodent intracerebral 50% – lethal doses per ml of blood – generally appearing about half way through the incubation period and persisting until death [49]). Most infectivity in rodent blood seemed to be associated with nucleated cells [50] – possibly B lymphocytes [51]. However, when human blood was spiked with large numbers of nucleated hamster-brain cells containing the scrapie agent and then fractionated, all components, including plasma and plasma derivatives contained infectivity, although some fractions – especially albumin – had only miniscule amounts [52] (Table 5).

As a precaution, the FDA took a very conservative position on risk of TSE transmission by human blood, and, in memoranda, CBER recommended deferral of donors at increased risk of developing CJD (i.e., donors with a convincing family history of familial CJD, those who might have received a dural allograft, and those who might have been treated with human cadaveric pituitary hormones) and withdrawal of products prepared from pooled plasma to which such donors, as well as donors later diagnosed with CJD, had inadvertently contributed [53].

However, very reassuring data have continued to accumulate from an increasing number of epidemiological studies, none suggesting that exposure to human blood or

Table 5: Distribution of TSE infectivity[a] in blood, blood components and plasma fractions.

Product	Total infectivity[b]	% Original infectivity recovered
Whole blood	10.0	100%
RBC	9.9	22%
Leucocytes, platelets	8.8	7%
Plasma	8.5	3%
Plasma	8.2	100%
Cryoprecipitate	6.8	0.7%
IgG	3.9	0.006%
Albumin	2.7	0.0004%

[a] Data are summarized from Brown et al [52] with permission. Human blood was spiked with viable nucleated cells from the trypsinized brain of a hamster infected with scrapie agent (263K strain). The blood was then fractionated by a scaled-down version of a conventional preparatory method.

[b] Infectivity (log10LD50) was assayed in hamsters (inoculated intracerebrally).

blood products constituted a risk factor for CJD [54, 55]. These studies were recently reviewed and analyzed by Schonberger and colleagues at the CDC [24, 25, 55]; their reports formed the basis for a recent reevaluation of policies of the U.S. Department of Health and Human Services and are briefly summarized. There have been no case reports of CJD convincingly attributable to blood exposure. Five published case-control studies failed to show any increased risk for CJD among people with a history of receiving blood or blood products [56-58]. Among 4,164 cases of CJD diagnosed on U.S. death certificates during an 18-year period from January 1979 to December 1996, the CDC (U.S. National Mortality Surveillance [55]) found no increase in annual death rates of CJD over the interval (in spite of presumed increased exposure to blood products in the U.S. population), no concomitant diagnosis of hemophilia, thalassemia or sickle-cell anemia – conditions treated repeatedly with human blood components or plasma derivatives – in any person dying with CJD. There was no diagnosis of CJD in any person less than 19 years old, despite the fact that, during the period studied, more than 38 million U.S. children were immunized – most in the second year of life – with a vaccine containing human albumin [25]. In a prospective study, initiated by the American Red Cross and continued by the National Blood Donor Resource Center [59], no case of CJD was recognized among 196 recipients of blood components from 15 donors who later came down with documented CJD; forty-two of the recipients survived for more than five years. In a similar larger study in Europe, no CJD was found in any recipient of a CJD-implicated blood component, although 13 recipients lived for more than ten years and eight for more than 15 years after transfusion [25]. In a survey of 120 Hemophilia Treatment Centers coordinated by the CDC [25, 60], no case of CJD has been recognized among more than 12,000 recipients of cryoprecipitate derived from pooled human plasma, and none of 30 patients who died had histopathological evidence of a TSE at autopsy. (Mean age of those subjects dying was 39 years, which should have been ample time for at least a few of them to have developed CJD, were they destined to do so).

So, although there seemed to be a remote theoretical risk that human blood or blood products prepared from persons incubating a TSE could transmit disease to a recipient, there was no evidence to suggest that such a thing had actually occurred. Also, on theoretical grounds, withdrawals of CJD-implicated plasma derivatives seemed unlikely to reduce substantially the very high theoretical probability that one or more donors contributing to a large pool of plasma (some from more than 60,000 individuals) would eventually come down with sporadic CJD [61, 62]. No screening question can identify those donors or pre-morbid laboratory test detect them. Finally, withdrawals generally come too late to retrieve most CJD-implicated products.

In recent years, some useful products derived from pooled human plasma – particularly immunoglobulin intravenous – have sometimes been in short supply, and withdrawals of CJD-implicated lots are thought to contribute to shortages. (Issues unrelated to CJD also played a role.) Responding to reports of such shortages, the Public Health Service (PHS) Advisory Committee on Blood Safety and Availability (ACBSA) reviewed the situation in January 1998 and advised the FDA to «...work with industry and appropriate consumer groups to relax current CJD guidelines on retrieval and withdrawal of blood products to the extent necessary to relieve shortages [of affected plasma derivatives]».

The PHS Blood Safety Committee, a senior executive committee of the U.S. Department of Health and Human Services including representatives of the CDC and National Institutes of Health as well as the FDA, again reviewed the issue. On 23 August 1998 the Surgeon General of the United States, Dr. David Satcher, announced a revision of policy consisting of four points: (i) Plasma derivatives should be withdrawn only if a donor developed new-variant CJD. (There are at least two reasons impelling the PHS to maintain a more conservative precautionary withdrawal policy for nvCJD than that adopted for sporadic, iatrogenic and familial forms of human TSE. (a) New-variant CJD has never been diagnosed in the U.S.A. and presumably represents a novel human infection with an exotic animal TSE agent strain. (b) The pathogenesis of nvCJD has not yet been well studied and may differ from that of other forms of CJD. The finding of large amounts of protease-resistant prion protein in tonsils of nvCJD patients at autopsy [63] and in the appendix of one nvCJD patient several years before onset of nvCJD [64] suggest that there may be a higher level of infectivity in lymphoid tissues and hence a greater risk of infectivity in blood than is the case for other forms of CJD (in which, as noted, infectivity has been detected in lymphoid tissues); (ii) Stop withdrawals of plasma derivatives for classical CJD and for CJD risk factors; (iii) Consider revised labelling of plasma derivatives; (iv) Continue to review epidemiological surveillance and laboratory studies relevant to the issue.

On 8 September 1998, CBER issued an interim change to previous guidance [53]: (i) «consistent with the procedures contained in the [previous CBER] memorandum, plasma derivatives should be retrieved, quarantined, destroyed and consignees notified only in the event that in-date products were manufactured from a donor who developed new variant CJD»; (ii) other provisions of the previous memorandum still represent FDA's current policy: (a) FDA recommends no change in policies concerning donor screening and deferral. (b) FDA continues to recommend withdrawal of CJD-implicated blood and blood components, including source plasma and leukocytes. (c) FDA recommends no change concerning notification of consignees or labelling (pending a decision by the PHS about changes in labelling appropriate for human plasma derivatives); (iii) CBER will develop and issue unified guidance on the issue.

Newer Important Issues Regarding TSEs

Several issues regarding TSEs must soon be addressed by regulatory authorities responsible for maintaining the safety of biological products. (i) How reliable are methods to detect TSE agents? (a) How sensitive and specific are assays of infectivity based on transmission of disease to animals of various species [66]? (b) How sensitive and reliable are assays based on detection of the protease-resistant prion proteins (PrP-res or PrP^{Sc}) that accumulate in brain [1, 67] (less consistently detected in lymphoid tissues [63, 64, 68, 69]) late in the incubation of TSEs? Does failure to detect PrP-res provide sufficient assurance of purity of source material or final product? How do sensitivity and specificity of the several immunodiagnostic tests described to detect PrP-res compare? Can the sensitivity of tests for abnormal PrP be increased so that they become reliably predictive of infectivity? (ii) Do «surrogate assays» for substances other than PrP-res that appear during the course of TSEs, like the 14-3-3 protein that can be detected in the CSF of patients with CJD and animals with scrapie during clinical illness [70-72], offer any assurance in determining purity of source? (iii) Are current manufacturing practices adequate to assure complete elimination of TSE agents from potentially contaminated facilities and equipment? Under carefully controlled conditions, TSE agents suspended in aqueous media can be inactivated down to undetectable levels by heat and by several chemical treatments [73, 74]; such treatments are generally recommended to clean potentially contaminated production facilities and equipment. However, it appears that such treatments may not reliably sterilize surfaces on which contaminated tissues are fixed or dried [41, 75-78]. Regulatory authorities should encourage development and validation of effective and practical methods for clean-up and disinfection; (iv) How reliable are «quantitative» assessments of the risk that regulated products may expose humans or animals to TSE agents [79, 80]? Regulatory authorities should increase their expertise in conducting and evaluating such risk assessments; (v) What implications, if any, have recently recognized TSEs of non-bovine animals for human health? A new TSE of animals has been recognized in the U.S.A., where chronic wasting disease – a scrapie-like syndrome – affects both captive and wild deer and elk [81]. BSE has never been detected in sheep, but sheep are susceptible to experimental infection with the BSE agent [82], and anxiety was publicly expressed recently about the possibility that sheep in Europe may have become accidentally infected [83]. Authorities must soon consider whether these newly recognized or postulated TSEs of animals pose any potential risk for the safety of regulated products and how to respond appropriately. To reach sound regulatory decisions concerning these and other issues regarding TSEs, it would be helpful to have available standard reference biological reagents – such as panels of well characterized materials containing various TSE agents – and harmonized methods for the diagnosis of TSEs and detection of the TSE agents.

References

1 Prusiner SB: Prion diseases and the BSE crisis. Science 1997;278:245-251.

2 United States Department of Health and Human Services, Food and Drug Administration. The Food and Drug Administration's development, issuance, and use of guidance documents. Federal Register 1997;62:8961-8970.

3 Will RG, Ironside JW, Zeidler M, Cousens SN, Estibeiro K, Alperovitch A, Poser S, Pocchiari M, Hofman A, Smith PG: A new variant of Creutzfeldt-Jakob disease in the UK. Lancet 1996; 347:921-925.

4 United States Department of Agriculture, Animal and Plant Health Inspection Service. Restrictions on the importation of ruminants, meat and meat products from ruminants, and certain other ruminant products (9 CFR Parts 94 and 96/Docket N° 97-127-1). Federal Register 1998;63:406-408.

5 United States Department of Health and Human Services, Food and Drug Administration. Substances prohibited from use in animal food or feed; animal proteins prohibited in ruminant feed; final rule. Federal Register 1997;62:30936-30978.

6 Brown P, Preece MA, Will RG: Friendly fire in medicine – hormones, homografts, and Creutzfeldt-Jakob disease. Lancet 1992;340:24-27.

7 United States Department of Health and Human Services, Centers for Disease Control and Prevention. Update: Creutzfeldt-Jakob disease in a second patient who received a cadaveric dura mater graft. MMWR 1989;38(3):37-38, 43.

8 Thadani V, Penar PL, Partington J, Kalb R, Janssen R, Schonberger LB, Rabkin CS, Prichard JW: Creutzfeldt-Jakob disease probably acquired from a cadaveric dura mater graft. Case report. J Neurosurg 1988;69:766-769.

9 United States Department of Health and Human Services, Centers for Disease Control and Prevention. Creutzfeldt-Jakob disease associated with cadaveric dura mater grafts – Japan, January 1979-May 1996. JAMA 1998;279:11-12.

10 United States Department of Health and Human Services Centers for Disease Control and Prevention. Creutzfeldt-Jakob disease associated with cadaveric dura mater grafts – Japan, January 1979-May 1996. MMWR 1997;46:1066-1069.

11 Duffy P, Collins G, Devoe AG, Streeten B, Cohen D: Possible person-to-person transmission of Creutzfeldt-Jakob disease. New Engl J Med 1974;290:693.

12 Fradkin JE, Schonberger LB, Mills JL, Gunn WJ, Piper JM, Wysowski DK, Thomson R, Durako S, Brown P: Creutzfeldt-Jakob disease in pituitary growth hormone recipients in the United States. JAMA 1991;265:880-884.

13 Gibbs CJ Jr, Asher DM, Brown PW, Fradkin JE, Gajdusek DC: Creutzfeldt-Jakob disease infectivity of growth hormone derived from human pituitary glands. New Engl J Med 1993;328: 358-359.

14 Gibbs CJ Jr, Asher DM, Kobrine A, Amyx HL, Sulima MP, Gajdusek DC: Transmission of Creutzfeldt-Jakob disease to a chimpanzee by electrodes contaminated during neurosurgery. J Neurol Neurosurg Psychiatry 1994;57:757-758.

15 Bernoulli C, Siegfried J, Baumgartner G, Regli F, Rabinowics T, Gajdusek DC, Gibbs CJ Jr: Danger of accidental person-to-person transmission of Creutzfeldt-Jakob disease by surgery. Lancet 1977;i:478-479.

16 Nevin S, McMenemy WH, Behrman D, Jones DP: Subacute spongiform encephalopathy: a subacute form of encephalopathy attributable to vascular dysfunction (spongiform cerebral atrophy). Brain 1960;83:519-564.

17 Klitzman RL, Alpers MP, Gajdusek DC: The natural incubation period of kuru and the episodes of transmission in three clusters of patients. Neuroepidemiology 1984;3:3-20.

18 Gajdusek D: Unconventional viruses and the origin and disappearance of kuru. Science 1977; 197:943-960.

19 Liberski PP, Gajdusek DC: Kuru: forty years later, a historical note. Brain Pathol 1997;7:555-560.

20 Monath TP, Heinz FX: Flaviviruses, in Fields BN, Knipe DM, Howley PM, Chanock RM, Melnick JL, Monath TP, Roizman B, Straus SE (eds): Fields Virology. 3rd ed Philadelphia, Lippincott-Raven Publishers, 1996, vol 1, pp 961-1034.

21 Gordon WS: Advances in veterinary research. Louping-ill, tick-borne fever and scrapie. Veterinary Record 1946;58:516-520.

22 Cuillé J, Chelle P-L: La tremblante du mouton est-elle inoculable? Comptes Rendus de l' Academie des Sciences 1936;203:1552-1554.

23 Detwiler L: USDA BSE surveillance program. NMHCC Conference. San Diego, California, U.S.A., 1998.

24 Schonberger LB: New variant Creutzfeldt-Jakob disease and bovine spongiform encephalopathy. Infect Dis Clin North Am 1998;12:111-121.

25 Schonberger LB, Belay ED: Summary of the clinicopathological features of transmissible spongiform encephalopathies (TSEs) in humans and Creutzfeldt-Jakob disease and blood safety data presented to PHS Advisory Committee on Blood Safety and Availability, January 28, 1998, with some updates through June 1998. (Available from the National Center for Infectious Diseases, Centers for Disease Control and Prevention) 1998.

26 United Kingdom Ministry of Agriculture Fisheries and Food. The decline in the BSE epidemic in Great Britain. BSE Enforcement Bulletin 1998(27):3-4.

27 Zeidler M, Stewart GE, Barraclough CR, Bateman DE, Bates D, Burn DJ, Colchester AC, Durward W, Fletcher NA, Hawkins SA, Mackenzie JM, Will RG: New variant Creutzfeldt-Jakob disease: neurological features and diagnostic tests. Lancet 1997;350(9082):903-907.

28 Zeidler M, Johnstone EC, Bamber RW, Dickens CM, Fisher CJ, Francis AF, Goldbeck R, Higgo R, Johnson-Sabine EC, Lodge GJ, McGarry P, Mitchell S, Tarlo L, Turner M, Ryley P, Will RG: New variant Creutzfeldt-Jakob disease: psychiatric features. Lancet 1997;350:908-910.

29 Will RG: New variant Creutzfeldt-Jakob disease, in Brown F, Griffiths E, Horaud F, Petricciani JC (eds): Safety of Biological Products Prepared from Mammalian Cell Culture. Dev Biol Stand. Basel, Karger, 1998, vol 93, pp 79-84.

30 Pearson GR, Wyatt JM, Henderson JPG-J T. J. Feline spongiform encephalopathy: a review. Veterinary Annual 1993;33:1-10.

31 United States Department of Health and Human Services, Food and Drug Administration. Substances prohibited from use in animal food or feed: protein derived from ruminants prohibited in ruminant feed. Advance notice of proposed rulemaking. 21 CFR 589. Federal Register 1996;61:24253-24255.

32 United States Department of Agriculture, Animal and Plant Health Inspection Service. Importation of animal products and byproducts from countries where BSE exists. Final rule (9 CFR Parts 94 and 95/Docket No 91-04). Federal Register 1991;56:63865.

33 United States Department of Agriculture, Animal and Plant Health Inspection Service. Importation of gelatin. Final Rule (9 CFR Part 94/Docket No 92-136-2). Federal Register 1993; 58:50250.

34 United States Department of Health and Human Services, Administration, Food and Drug Administration. Bovine-derived materials; agency letters to manufacturers of FDA-regulated products. Federal Register 1994;59:44591-44594.

35 United States Department of Health and Human Services, Administration, Food and Drug Administration. Bovine spongiform encephalopathy (BSE) in products for human use; guidance for industry on the sourcing and processing of gelatin to reduce potential risk; availability. Federal Register 1997;62:52345-52346.

36 Schrieber R, Seybold U: Gelatine production, the six steps to maximum safety, in Brown F (ed): Transmissible Spongiform Encephalopathies – Impact on Animal and Human Health. Dev Biol Stand. Basel, Karger, 1993, vol 80, pp 195-198.

37 Schrieber R: Validation study of the clearance of scrapie during gelatin processing. Meeting of the FDA Transmissible Spongiform Encephalopathies Advisory Committee 23 April 1997. Bethesda, Maryland, U.S.A.: FDA United States Department of Health and Human Services 1997:182-204.

38 Kimberlin RH, Walker CA: Pathogenesis of experimental scrapie, in Bock G, Marsh J (eds): Novel Infectious Agents and the Central Nervous System. Ciba Foundation Symposium. Chichester, Wiley and Sons, 1988, vol 135, pp 37-62.

39 Office International des Epizooties. Bovine spongiform encephalopathy (Chapter 3.2.13). International Animal Health Code, Special Edition. Paris: OIE, 1997:267-274.

40 Green C: Overview of U.S. oleochemical industry and production processes. FDA Transmissible Spongiform Encephalopathies Advisory Committee Meeting 15-16 April 1998. Bethesda, Maryland, U.S.A.: United States Department of Health and Human Services, FDA, 1998:159-187.

41 Taylor DM, Woodgate SL, Atkinson MJ: Inactivation of the bovine spongiform encephalopathy agent by rendering procedures. Vet Rec 1995;137:605-610.

42 Asher DM: Bovine sera used in the manufacture of biologicals: current concerns and policies of the U.S. Food and Drug Administration concerning the transmissible spongiform encephalopathies, in Brown F, Cartwright T, Horaud F (eds): Animal Sera, Animal Sera Derivatives and Substitutes Used in the Manufacturing of Pharmaceuticals: Viral Safety and Regulatory Aspects. Dev Biol Stand. Basel, Karger, 1999, vol 99, pp 41-44.

43 Bradley R: Implication of new BSE data on gelatin and UK action. FDA Transmissible Spongiform Encephalopathies Advisory Committee Meeting 15-16 April 1998. Bethesda, Maryland, U.S.A.: United States Department of Health and Human Services, FDA, 1998:157-165.

44 Hadlow W, Race R, Kennedy R, Eklund C: Natural infection of sheep with scrapie virus, in Prusiner S, Hadlow W (eds): Slow Transmissible Diseases of the Nervous System. New York, Academic Press, 1979, vol 2.

45 Brown P, Gibbs CJ, Jr, Rodgers-Johnson P, Asher DM, Sulima MP, Bacote A, Goldfarb LG, Gajdusek DC: Human spongiform encephalopathy: the NIH series of 300 cases of experimentally transmitted disease. Ann Neurol 1994;35:513-529.

46 Diringer H: Sustained viraemia in experimental hamster scrapie. Arch Virol 1984;82:105-109.

47 Manuelidis EE, Gorgacz EJ, Manuelidis L: Viremia in experimental Creutzfeldt-Jakob disease. Science 1978;200:1069-1071.

48 Manuelidis EE, Kim JH, Mericangas JR, Manuelidis L: Transmission to animals of Creutzfeldt-Jakob disease from human blood. Lancet 1985;2:896-897.

49 Casaccia P, Ladogana A, Xi XG, Pocchiari M: Levels of infectivity in the blood throughout the incubation period of hamsters peripherally injected with scrapie. Arch Virol 1989;108:145-149.

50 Kuroda Y, Gibbs CJ Jr, Amyx HL, Gajdusek DC: Creutzfeldt-Jakob disease in mice: persistent viremia and preferential replication of virus in low-density lymphocytes. Infect Immun 1983; 41:154-161.

51 Klein MA, Frigg R, Flechsig E, Raeber AJ, Kalinke U, Bluethmann H, Bootz F, Suter M, Zinkernagel RM, Aguzzi A: A crucial role for B cells in neuroinvasive scrapie. Nature 1997;390: 687-690.

52 Brown P, Rohwer RG, Dunstan BC, MacAuley C, Gajdusek DC, Drohan WN: The distribution of infectivity in blood components and plasma derivatives in experimental models of transmissible spongiform encephalopathy. Transfusion 1998;38:810-816.

53 United States Department of Health and Human Services, Food and Drug Administration. Revised precautionary measures to reduce the possible transmission of Creutzfeldt-Jakob disease (CJD) by blood and blood products; guidance document; availability [Docket No. 97D-0316]. Federal Register 1997;62(184):49694-49695.

54 Ricketts MN, Cashman NR, Stratton EE, Elsaadany S: Is Creutzfeldt-Jakob disease transmitted in blood? Emerg Infect Dis 1997;3:155-163.

55 Holman RC, Khan AS, Belay ED, Schonberger LB: Creutzfeldt-Jakob disease in the United States, 1979-1994: using national mortality data to assess the possible occurrence of variant cases. Emerg Infect Dis 1996;2:333-337.

56 Esmonde TFG, Will RG, Slattery JM, Knight R, Harries Jones R, Desilva R, Matthews WB: Creutzfeldt-Jakob Disease and Blood Transfusion. Lancet 1993;341:205-207.

57 Wientjens DP, Davanipour Z, Hofman A, Kondo K, Matthews WB, Will RG, van Duijn CM: Risk factors for Creutzfeldt-Jakob disease: a reanalysis of case-control studies. Neurology 1996; 46:1287-1291.

58 van Duijn CM, Delasnerie-Laupretre N, Masullo C, Zerr I, de Silva R, Wientjens DP, Brandel JP, Weber T, Bonavita V, Zeidler M, Alperovitch A, Poser S, Granieri E, Hofman A, Will RG: Case-control study of risk factors of Creutzfeldt-Jakob disease in Europe during 1993-95. European Union (EU) Collaborative Study. Lancet 1998;351:1081-1085.

59 Sullivan M: Epidemiological studies of CJD and blood, blood components and derivatives. FDA Transmissible Spongiform Encephalopathies Advisory Committee Meeting 6-7 October 1997. Bethesda, Maryland, U.S.A.: United States Department of Health and Human Services, FDA, 1997:92-98.

60 Souci M: Epidemiological studies of CJD and blood, blood components and derivatives. FDA Transmissible Spongiform Encephalopathies Advisory Committee Meeting 6-7 October 1997. Bethesda, Maryland, U.S.A.: United States Department of Health and Human Services, FDA, 1997:98-105.

61 Lynch TJ, Weinstein MJ, Tankersley DL, Fratantoni JC, Finlayson JS: Considerations of pool size in the manufacture of plasma derivatives. Transfusion 1996;36:770-775.

62 Brown P: Donor pool size and the risk of blood-borne Creutzfeldt-Jakob disease. Transfusion 1998;38:312-315.

63 Hill AF, Zeidler M, Ironside J, Collinge J: Diagnosis of new variant Creutzfeldt-Jakob disease by tonsil biopsy. Lancet 1997;349:99-100.

64 Hilton DA, Fathers E, Edwards P, Ironside JW, Zajicek J: Prion immunoreactivity in appendix before clinical onset of variant Creutzfeldt-Jakob disease. Lancet 1998;352:703-704.

65 Asher DM: Transmission of human spongiform encephalopathies to experimental animals: comparison of the chimpanzee and squirrel monkey, in Brown F (ed): Transmissible Spongiform Encephalopathies. Impact on Animal and Human Health. Dev Biol Stand. Basel, Karger, 1993, vol 80, pp 9-13.

66 Race RE, Priola SA, Bessen RA, Ernst D, Dockter J, Rall GF, Mucke L, Chesebro B, Oldstone MB: Neuron-specific expression of a hamster prion protein minigene in transgenic mice induces susceptibility to hamster scrapie agent. Neuron 1995;15:1183-1191.

67 Safar J, Wille H, Itri V, Groth D, Servan H, Torchia M, Cohen FE, Prusiner SB: Eight prion strains have PrPSc molecules with different conformations. Nature Medicine 1998;4:1157-1165.

68 Race R, Jenny A, Sutton D: Scrapie infectivity and proteinase K-resistant prion protein in sheep placenta, brain, spleen, and lymph node: implications for transmission and antemortem diagnosis. J Infect Dis 1998;178:949-953.

69 van Keulen LJ, Schreuder BE, Meloen RH, Mooij-Harkes G, Vromans ME, Langeveld JP: Immunohistochemical detection of prion protein in lymphoid tissues of sheep with natural scrapie. J Clin Microbiol 1996;34:1228-1231.

70 Harrington MG, Merril CR, Asher DM, Gajdusek DC: Abnormal proteins in the cerebrospinal fluid of patients with Creutzfeldt-Jakob disease. New Engl J Med 1986;315:279-283.

71 Hsich G, Kenney K, Gibbs CJ, Lee KH, Harrington MG: The 14-3-3 brain protein in cerebrospinal fluid as a marker for transmissible spongiform encephalopathies. N Engl J Med 1996; 335:924-930.

72 Will RG, Zeidler M, Brown P, Harrington M, Lee KH, Kenney KL: Cerebrospinal-fluid test for new-variant Creutzfeldt-Jakob disease. Lancet 1996;348:955.

73 Rohwer RG: Scrapie infectious agent is virus-like in size and susceptibility to inactivation. Nature 1984;308:658-662.

74 Rohwer RG: Virus-like sensitivity of scrapie agent to heat inactivation. Science 1984;223:600-602.

75 Taylor D, McConnell I: Autoclaving does not decontaminate formol-fixed scrapie tissues. Lancet 1988;1:1463-1464.

76 Taylor DM: Inactivation of SE Agents. Br Med Bull 1993;49:810-821.

77 Taylor DM, Fraser H, McConnell I, Brown DA, Brown KL, Lamza KA, Smith GRA: Decontamination studies with the agents of bovine spongiform encephalopathy and scrapie. Arch Virol 1994;139:313-326.

78 Taylor DM, Fernie K: Exposure to autoclaving or sodium hydroxide extends the dose-response curve of the 263K strain of scrapie agent in hamsters. J Gen Virol 1996;77:811-813.

79 Bader F, Davis G, Dinowitz M, Garfinkle B, Harver J, Kozak R, Lubiniecki A, Rubino B, Schubert D, Wiebe M, Woollett G: Assessment of risk of bovine spongiform encephalopathy in pharmaceutical products. Part 1. BioPharm 1998;11:20, 22, 24-26, 28-31, 56.

80 Bader F, Davis G, Dinowitz M, Garfinkle B, Harver J, Kozak R, Lubiniecki A, Rubino B, Schubert D, Wiebe M, Woollett G: Assessment of risk of bovine spongiform encephalopathy in pharmaceutical products. Part 2. Example risk assessment for a hypothetical product. BioPharm 1998;11:18-20, 22, 24-30.

81 Spraker TR, Miller MW, Williams ES, Getzy DM, Adrian WJ, Schoonveld GG, Spowart RA, O'Rourke KI, Miller JM, Merz PA: Spongiform encephalopathy in free-ranging mule deer (Odocoileus hemionus), white-tailed deer (Odocoileus virginianus) and Rocky Mountain elk (Cervus elaphus nelsoni) in northcentral Colorado. J Wildl Dis 1997;33:1-6.

82 Foster JD, Bruce M, McConnell I, Chree A, Fraser H: Detection of BSE infectivity in brain and spleen of experimentally infected sheep. Vet Rec 1996;138:546-548.

83 Butler D: Doubts over ability to monitor risks of BSE spread to sheep. Nature 1998;395:6-7.

84 Bruce ME, Will RG, Ironside JW, McConnell I, Drummond D, Suttie A, McCardle L, Chree A, Hope J, Birkett C, Cousens S, Fraser H, Bostock CJ: Transmissions to mice indicate that «new variant» CJD is caused by the BSE agent. Nature 1997;389:498-501.

85 Hill AF, Desbruslais M, Joiner S, Sidle KCL, Gowland I, Collinge J, Doey LJ, Lantos P: The same prion strain causes vCJD and BSE. Nature 1997;389:448-450.

Dr. D.M. Asher, Lab. of Method Development, Div. of Viral Products, OVRR/CBER/USFDA, HFM-470 FDA, 1401 Rockville Pike, Rockville, MD 20852-1448, USA.

Brown F, Griffiths E, Horaud F, Schild GC (eds): A Celebration of 50 Years of Progress in Biological Standardization and Control at WHO.
Dev Biol Stand. Basel, Karger, 1999, vol 100, pp 119-129

Development and Standardisation of DNA Vaccines

K. Cichutek

Department of Medical Biotechnology, Langen, Germany

INTRODUCTION

The first interesting example of in vivo DNA inoculation, as I remember it, was described in an abstract by William Haseltine in 1988. Plasmid DNA comprising the full-length genome of simian immunodeficiency virus had been inoculated intramuscularly (i.m.) into monkeys and, astonishingly, resulted in a chronic infection. Evidently, at least one functional copy of the DNA must have been taken up and expressed by a somatic cell in vivo. This result was regarded by scepticism because efficient uptake of DNA into cultured primate cells was known to be successful only if transfection reagents like cationic liposomes were used. In view of the possible risk of spontaneous DNA uptake by personnel working in gene laboratories, specific recommendations for handling of DNA were issued in Germany. In fact, this risk is minimal, if good microbiological techniques as practised in such facilities are used.

In 1990 the first paper describing the immunisation of mice following deliberate inoculation of non-complexed, so-called «naked» DNA, was published and raised considerable interest [22]. It was followed by a number of other publications describing the seemingly effortless possibility of transferring DNA in vivo into somatic cells of mice and non-human primates including chimpanzees followed by the induction of immune responses to the gene products encoded by the inoculated DNA [4, 5, 11, 15-17, 19, 22, 24]. Initially, not only DNA, but also RNA was used to inoculate animals. Therefore, the term nucleic acid vaccination was coined during an early WHO meeting in 1994 which was organised to review the development of the field. However, as DNA is more stable and at least as easy to use for in vivo inoculations as RNA, almost all scientists involved in the development of nucleic acid vaccines today are using DNA.

DEVELOPMENT OF DNA VACCINES FOR HUMAN USE

Nucleic acid vaccines, also termed genetic vaccines, contain non-complexed and non-vectored nucleic acids intended to be used as prophylactic vaccines in humans or animals. As DNA is more stable as RNA and can be readily amplified and prepared from recombinant *E. coli* bacteria, plasmid DNA comprising one or more synthetic mammalian genes is mostly used today. Consequently, most plasmid DNAs used for DNA vaccination comprise a bacterial origin of replication (Col E1) and do not contain any mammalian replication origin as this would result in the unwanted replication of the DNA in mammalian cells. The structure of a plasmid typically used for DNA vaccination is shown in Figure 1. The synthetic mammalian gene comprises a transcriptional promoter like the cytomegalovirus intron A promoter suitable for efficient gene expression in mammalian cells. The promoter is linked to a region encoding an immunogenic protein of the relevant pathogen preceded by a cap site necessary for correct mRNA initiation in mammalian cells. The expression gene is completed by a poly-adenylation (poly(A)) signal necessary to allow correct termination of the mRNA. The poly(A)-signals derived either from bovine growth hormone or from Simian Virus 40 are often used because they allow very little read-through transcription possibly resulting in the transcription of other proximal genes.

During a typical preclinical experiment, plasmid DNA which can be purified in mg amounts e.g. from a one litre culture of the respective recombinant *E. coli* strain is extracted and purified by affinity column chromatography (Fig. 2). Contaminating endotoxins are removed using commercially available kits. Endotoxins may result in reduced efficiency of immunisation, most probably by reducing the transfection efficiency in vivo. The DNA is dissolved in saline solution and, typically, 100 μg of DNA in volumes of 50 to 100 μl are injected into the *musculus femoralis* of a laboratory mouse. Some training in the injection technique is initially required to reproducibly achieve detectable immunisation of each animal injected. As an alternative to i.m. inoculation, intradermal DNA injection (not s.c. injection) and so-called gene gun or bolistic inoculation result in efficient immunisation against the

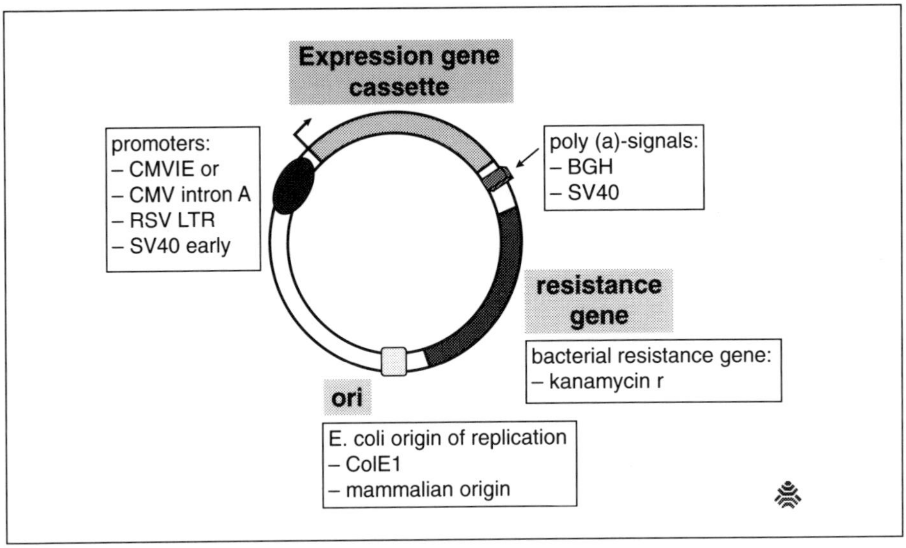

Fig. 1: Plasmid constructs used for DNA vaccination.

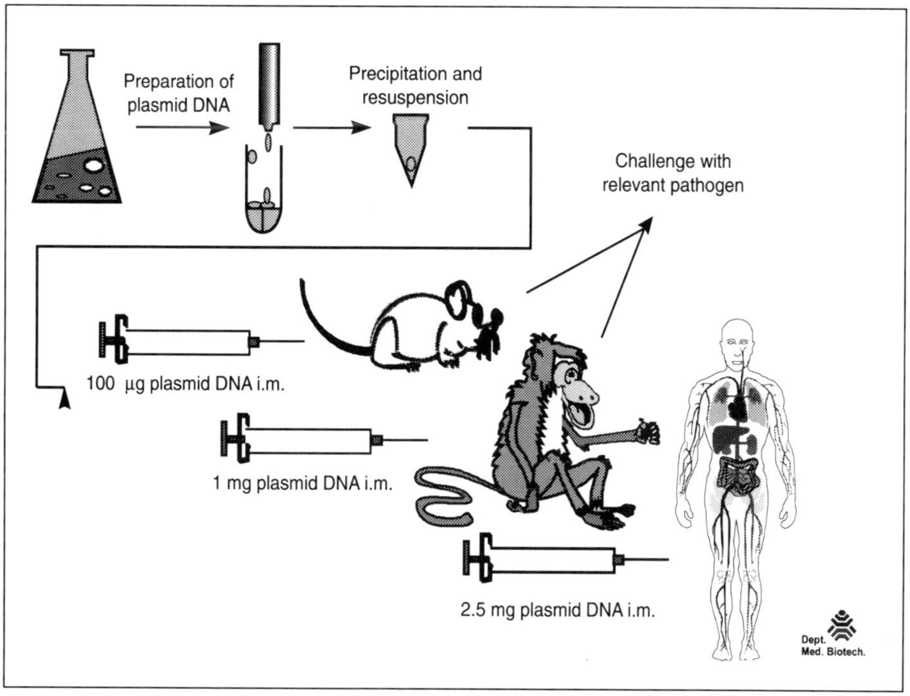

Fig. 2: Experimental design; preclinical studies.

antigen encoded by the plasmid used. For gene gun injections, 10 or 100 ng of DNA are coated onto gold particles which are bolistically delivered into the skin by pressured gas using the gene gun device. The gene gun as well as the use of mediators have been shown to increase the proportion of injected animals which seroconvert. For gene gun inoculations, at least 100-fold less DNA can be used to produce results comparable with those achieved by i.m. inoculations. For DNA inoculation in non-human primates including chimpanzees, mg amounts of DNA have to be used to achieve consistent results. Consequently, a comparable amount of DNA has also been used in clinical trials in human beings.

The i.m. inoculation of DNA leads to quite efficient spontaneous uptake by myocytes which then express the encoded gene. Initially, myocytes were suspected to serve as professional antigen presenting cells (APCs) because they also express MHC I, although at a low level. However, for efficient priming of the immune system, additional expression of co-stimulatory molecules by the antigen presenting cells is necessary. This may not be readily achieved by transfected muscle cells. As some of the DNA injected may be taken up into the blood stream and may lead to transfection of liver cells, antigen processing and presentation is possibly also performed by professional APCs in the liver, in lymph nodes and in other organs. The delivered genes are expressed in the somatic cells transfected and presented by the conventional pathways through the MHC I complex (Fig. 3). This is one of the important advantages of DNA vaccines compared to subunit and live attenuated vaccines. Similar to efficient immune stimulation during normal viral infections, for

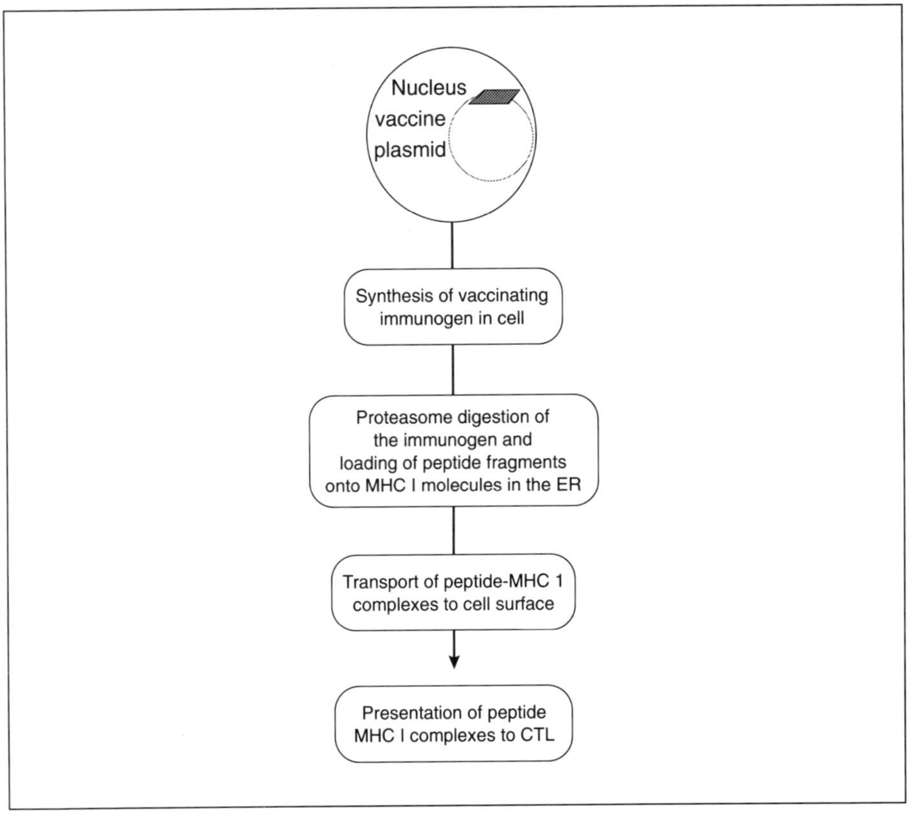

Fig. 3: MHC I, synthesis of immunogen in cell.

example, DNA vaccination results in antigen expression and presentation by somatic cells in vivo. This form of antigen presentation can only be achieved inefficiently using subunit vaccines. Similar to subunit vaccines, however, DNA vaccines do not carry the risk of transmission of the respective replication competent pathogen. Thus, DNA vaccines share the safety of subunit vaccines as only single recombinant antigens are used for vaccination in the absence of any risk of replication competent pathogen delivery to the vaccinee. In contrast to subunit vaccines, however, efficient MHC I presentation and priming of cellular immune responses is achieved by DNA vaccines. In addition to MHC I presentation, uptake by APCs of the antigen produced from transfected cells in vivo following DNA vaccination probably also results in MHC II presentation (Fig. 4).

Many investigators have shown that i.m. and i.d. DNA inoculation of mice results in a strong Th1 response characterised by the secretion of IFN-γ and IL-2 from stimulated immune cells [4-6, 11, 14, 17, 26]. In concordance, IgG2a antibodies are produced. A possible mechanism underlying this Th1 bias is the initially strong induction of IFN-γ secretion, probably due to unmethylated CpG nucleotide sequence motifs present in bacterial plasmid DNA. In fact, efficient immune stimulation has been associated with the presence of such CpG motifs in the inoculated

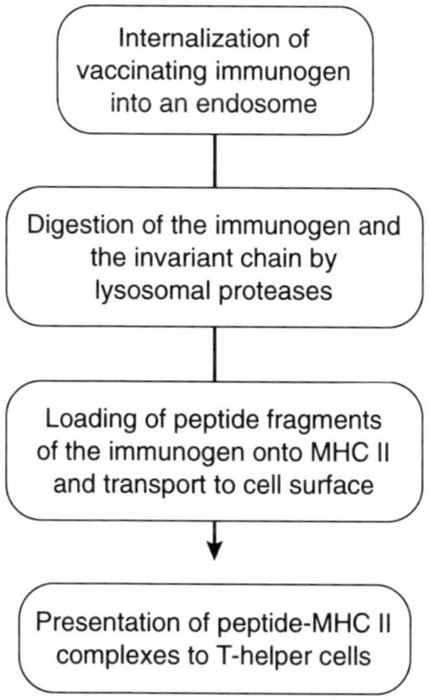

Fig. 4: MHC II, synthesis of immunogen in cells or internalization of immunogen from outside of cell.

plasmid DNA and they may contribute to the efficient immunisation by DNA inoculation similar to the effect of adjuvants in subunit vaccines.

Immunisation against a variety of viral, bacterial and parasitic pathogens following plasmid DNA inoculation of animals has been described (Figs 5, 6). Investigators have also been able to show protection against infection or disease in suitable animal models of infectious human disease. Interesting examples for DNA vaccine efficiency have included the demonstration of protection against heterologous strains of influenza A virus using a gene encoding the viral nucleoprotein (NP) [20] and the production of a vaccine against tuberculosis [7, 8] and malaria [23]. DNA vaccines are expected to be cost-effectively produced, easily manufactured and to show extended stability during transport and storage. In view of these characteristics, applications in less developed regions of the world may also be foreseen. Consequently, DNA vaccine development against malaria, tuberculosis and AIDS may probably be most desirable.

Due to some difficulties during delivery of DNA to non-human primates, which may be due to the more rigid nature of skin or muscle tissue compared to that of rodents, clinical trials in humans were initially met with some scepticism. However, in recent publications describing the result of clinical phase I trials evidence for efficient immunisation of humans following DNA inoculation has been presented [1, 4, 23]. The demonstration of MHC I-restricted CD8+ and peptide-specific cytotoxic T-lymphocytes following three inoculations of DNA encoding a liver stage-specific antigen derived from the parasite *Plasmodium falciparum* has been especially convincing. Thus, the development of DNA vaccines for human use has proceeded to the stage of successful clinical phase I or phase I/II trials.

- Influenza A virus
  - HA (NAb; protective), matrix protein, NP
- HIV-1
  - gp 120-SU, gp 160 (NAb)
- Bovine herpes virus I
  - gIV
- Rabies virus
  - surface glycoprotein (Nab; protective)
- Hepatitis C virus
  - core protein
- Hepatitis D virus
- Lymphocytic choriomeningitis virus (LCMV)
  - NP
- Cotton-tail rabbit papilloma virus (CRPV)
  - major capsid protein (NAb)
- Herpes simplex virus (HSV)
  - gB, gD (NAb; protective with BHV)

Fig. 5: Antibody production in animals against viral pathogens following i.m. inoculation of DNA.

- *Parasites*
- Schistozoma japonicum
- Leishmania major
  - protective
- Plasmodium yoelii
  - protective
- *Bacteria*
- Mycobacterium tuberculosis
  - protective
- Salmonella typhi
- Mycoplasma pulmonis (HSV)
  - protective

Fig. 6: Antibody production in animals against non-viral pathogens following i.m. inoculation of DNA.

STANDARDISATION OF DNA VACCINES FOR HUMAN USE

In 1994 the WHO organised a conference in Geneva to review the development of DNA vaccines together with the pharmaceutical industry, scientists working at publicly funded research institutions and regulatory scientists employed by federal higher authorities involved in the regulation and licensing of medicinal products [2, 16]. The FDA presented a first draft of the later published guideline on «polynucleotide vaccines». European regulatory scientists pointed out that DNA vaccines would fall under the definition of medicinal products for gene therapy for which a European Note for Guidance was already drafted [12]. During the meeting it became obvious that DNA vaccines could potentially be produced at relatively low costs. The manufacturing process of plasmid DNA from bacteria could be effectively standardised. DNA could be stored and transported at low costs and would have extended stability. In addition, animal experiments, at that time mostly performed in mice, showed very positive results, which promised the development of new vaccines against diseases for which no vaccines exist (like AIDS), better vaccines to improve the potential of the nowadays available vaccines (like the influenza virus vaccine) and to allow production of cheaper and more effective vaccines needed in less developed countries (like a vaccine against malaria or tuberculosis). In addition, the concept of developing guidelines for the standardisation of novel medicinal products in parallel with their scientific and industrial development was once again stressed. Thus, development of regulations would not be disaster-driven, as sometimes seen in the past, but would accompany product development. However, the fulfilment of this concept then led to the anticipation of risks and the discussion of merely theoretical risks for which no indication had been found during preclinical animal experiments or human clinical trials. Nevertheless, this seemed to be and still is especially important when DNA vaccines are considered for human use as vaccines are often used in millions of humans and also young children for prophylactic purposes only. This normally affords a very considerable benefit-risk ratio. In view of the greatest, but still theoretical risk of DNA vaccines discussed during the WHO and other succeeding conferences on DNA vaccine development and standardisation, namely the induction of malignant tumours as a consequence of chromosomal integration and insertional mutagenesis by the DNA delivered in vivo, it will be a major challenge to convince the public that DNA vaccines can still be used without any greater risk than other conventional vaccines.

Taking into account the «Guidelines Assuring the Quality of DNA Vaccines» published by WHO [18], there are three major topics to be considered during

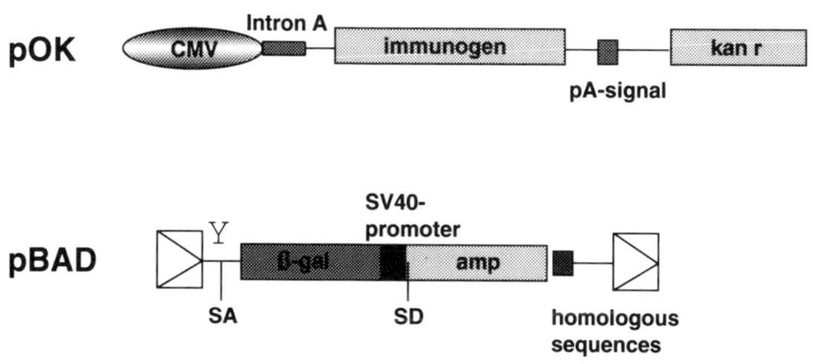

- design of expression genes
- absence of sequences mediating chromosomal integration of plasmid DNA
- translation of abberrant gene products

Fig. 7: Design of plasmid.

development and safety testing, i.e. chromosomal integration of the foreign DNA, induction of tolerance and induction of auto-immune disease. Klinman et al [9] adressed the latter two topics by experiments using a *Plasmodium falciparum* gene. He found no evidence for induction of or increase of severity of already present auto-immune disease. More specifically, the level of anti-DNA or anti-self antibodies did not increase following inoculation of plasmid DNA by several routes. In addition, tolerance to the antigen expressed was not evident in inoculated mice. Tolerance can be induced by DNA inoculation and other methods of immunisation in newborn mice which do not have a fully developed immune system during the first days after birth. The third topic, chromosomal integration of the plasmid DNA, was investigated following i.m. inoculation of DNA in mice [4, 13]. DNA was extracted and chromosomal DNA was separated from episomal DNA by agarose gel chromatography. Absence of integrated plasmid DNA was found with a level of detection of one copy in 10^6 cells. This indicates that in myocytes, which take up the inoculated DNA, the foreign DNA is not integrated into the chromosomal DNA or integrated at a very low level. More sensitive techniques will be available in the future to address this question also in the draining lymph nodes or other organs of the immune system. However, given that after transfection of mammalian cells in vitro, all cells integrate the foreign DNA even in the absence of selection for transfected marker genes. Chromosomal integration of plasmid DNA probably occurs in vivo and will be detected with more sensitive techniques. However, not every single process of integration will lead to insertional mutagenesis, meaning activation of an oncogene or functional deletion of a tumour suppressor gene. In addition, the induction of malignant tumours requires a series of genetic events which have to occur in addition to insertional mutagenesis. Taking these facts into account, estimates of the risk of cancer induction in healthy vaccinees have been made. A risk of one possible negative event in 10^6 vaccinees was hypothesized to be imposed by DNA vaccination [10], given that the assumptions made were realistic. Thus, the risk of tumour induction by DNA inoculation has to be taken into consideration, but at this point, seems to have a probability which allows for clinical trials in humans.

With a view to plasmid DNA construction, the expression genes should be carefully designed. Sequences possibly mediating chromosomal integration of the plasmid DNA should be avoided. In this regard, search for sequence homologies between the plasmid sequences and known human genomic sequences may help. Clearly, retroviral LTRs or sequences with homology to the packaging sequences of Foamy Retrovirus, lentiviruses or human endogenous retroviruses, as far as known, should be avoided. The presence in the plasmid of possible coding regions leading to the false expression of unwanted genes in mammalian cells is also not acceptable. Next, plasmid DNA intended to be used in human clinical trials has to be produced according or similar to GMP in Germany and probably also in most other European Union member states. However, a potentially slightly lower standard of production is sometimes discussed during scientific meetings, especially with a view to initial clinical trials of phase I/II with a very low number of terminally ill patients. The quality of the plasmid DNA should be defined and validated tests should be used to control these standards in each batch (Fig. 8). For example, the amount of bacterial DNA, endotoxins and aberrant plasmid structures in the preparations is relevant, also in view of the transfection efficiency to be achieved in vivo. As already mentioned, i.m. inoculation probably results in episomal uptake of the

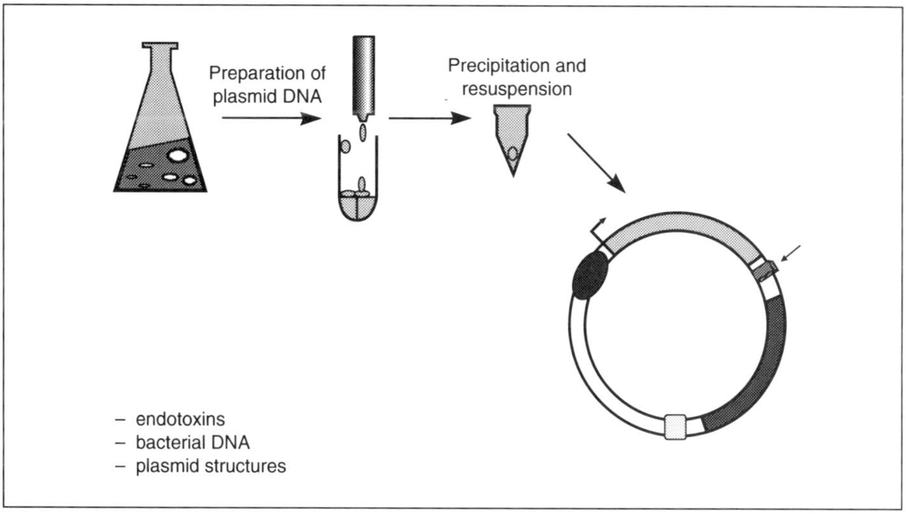

Fig. 8: Purification of nucleic acid.

DNA in muscle cells (Fig. 9). This should be verified with the best methods available for the specific plasmid DNA to be used in humans. In addition, absence of chromosomal integration of the inoculated DNA should be checked in the tissue inoculated, in draining lymph nodes and in other organs which are possible targets of in vivo DNA migration like the spleen and the gonads. In addition, distribution

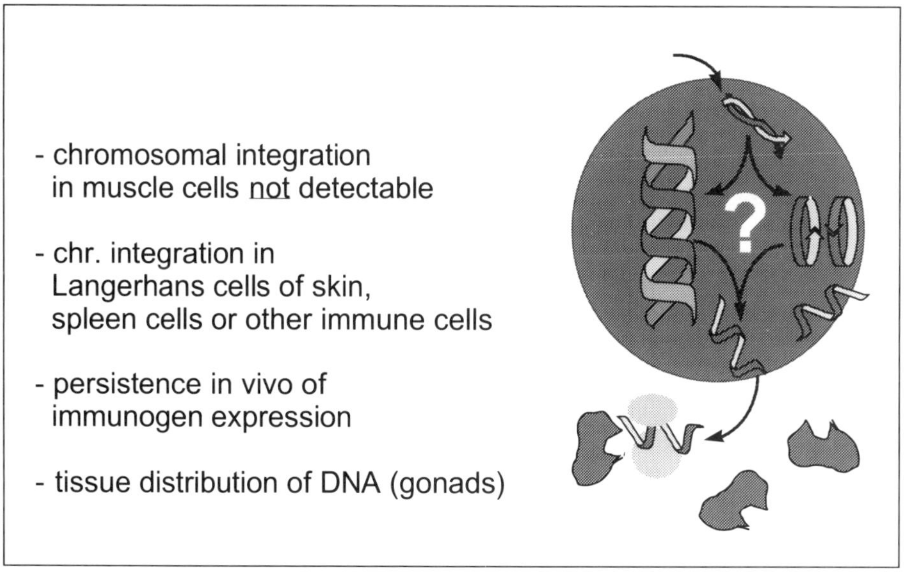

Fig. 9: Intra- and extra-cellular fate of DNA.

- Plasmid DNA unable to induce or accelerate the development of autoimmune disease.
- Certain organ-specific autoimmune diseases may be facilitated with some plasmid DNAs.
- Tolerance may be induced in some populations.
- DNA vaccines may be ineffective under certain circumstances.
- Chromosomal integration and tissue distribution remain issues.

Fig. 10: Risk assessment and efficacy of DNA vaccines.

of the plasmid DNA to other tissues should be assessed. Where applicable, so-called «worst case scenarios» like the i.v. inoculation of DNA and the resulting tissue distribution in vivo may be addressed by preclinical experiments. Another major topic would be the possible induction of tolerance. Any information indicating possible tolerance induction in animals should be assembled. Depending on the vaccine to be developed, the persistence of antigen expression in vivo following the route of inoculation to be tested in humans should be analysed. Generally, a suitable animal model relevant for assessing the potential risks has to be chosen. A case-by-case evaluation of the pharmacological-toxicological data will leave room for a flexible approach.

References

1 Calarota S, Bratt G, Nordlund S, Hinkula J, Leandersson AC, Sandstrom E, Wahren B: Cellular cytotoxic response induced by DNA vaccination in HIV-1-infected patients. Lancet 1998; 351(9112):1320-1325.

2 Cichutek K: Nucleic Acid vaccines: A prophylactic gene therapy? Vaccines 1994;12:1520-1525.

3 Cichutek K: Obtaining marketing authorization for nucleic acid vaccines in the European-Union. Ann NY Acad Sci 1995;772:178-185.

4 Donnelly JJ, Ulmer JB, Shiver JW, Liu M: DNA vaccines. Annu Rev Immunol 1997;15:617-648.

5 Food and Drug Administration, Center of Biologics Evaluation and Research, Office of Vaccine Research and Review, Points to consider on plasmid DNA vaccines for preventive infectious disease indications. Docket No. 96N-0400, 1996.

6 Fynan EF, Webster RG, Fuller DH, Haynes JR, Santoro JC, Robinson HL: DNA vaccines: Protective immunizations by parenteral, mucosal, and gene-gun inoculations. Proc Natl Acad Sci USA 1993;90:11478-11482.

7 Huygen K, Content J, Denis O, Montgomery DL, Yawman AM, Deck RR, DeWitt CM, Orme IM, Baldwin S, D'Souza CS, Drowart A, Lozes E, Vandenbussche P, Mooren J-P, Liu MA, Ulmer JB: Immunogenicity and Protective efficacy of a tuberculosis DNA vaccine. Nature Medicine 1996;2:893-898.

8 Tacson RE, Colston MJ, Ragno S, Stavropoulos E, Gregory D, Lowrie DB: Vaccination against tuberculosis by DNA injection. Nature Med 1996;2:888-92.

9 Klinman DM, Takeno M, Ichino M, Gu M, Yamshchikov G, Mor G, Conover J: DNA vaccines: safety and efficacy issues. Springer Semin. Immunopathol 1997;19:245-256.

10 Kurth R: Risk potential of the chromosomal insertion of foreign DNA. Ann NY Acad Sci 1995;772:140-151.

11 Mölling K: Nackte DNA zur Impfung und Therapie. DMW 997;23:757-760.

12 NfG: Gene therapy product quality aspects in the production of vectors and genetically modified somatic cells. CPMP/BWP/1106/97.

13 Nichols WW, Ledwith BJ, Manam SV, Toilo PJ: Potential DNA vaccine integration into host cell genome. Ann NY Acad Sci 1995;772:30-39.

14 Raz E, Carson DA, Parker SE, Parr TB, Abai AM, Aichinger G, Gromkowski H, Singh M, Lew D, Yankauckas MA, Baird SM, Rhodes GH: Intradermal gene immunization: The possible role of DNA uptake in the induction of cellular immunity to viruses. Proc Natl Acad Sci USA 1994; 91:9519-9523.

15 Robinson HL, Torres CAT: DNA vaccines. Seminars in Immunology 1997;9:271-283.

16 Spier RE: Nucleic Acid Vaccines. Vaccine 1995;13(1):131-132.

17 Tang D, De Vit M, Johnston SA: Genetic immunization is a simple method for eliciting an immune response. Nature 1992;356:152-154.

18 Technical Report Series No. 17 January 1997.

19 Tighe H, Corr M, Roman M, Raz E: Gene vaccination: plasmid DNA is more than just a blueprint. Immunology Today 1998;19(2):89-97.

20 Ulmer JB, Donnelly JJ, Parker SE, Rhodes GH, Felgner PL, Dwarki VJ, Gromkowski SH, Deck RR, DeWitt CM, Friedman A, Hawe LA, Leander KR, Martinez D, Perry HC, Shiver JW, Montgomery DL, Liu M: Heterologous Protection Against Influenza by Injection of DNA Encoding a Viral Protein. Science 1993;259:1745-1749.

21 Vahlsing HL, Yankauckas MA, Sawdey M, Gromkowski SH, Manthorpe M: Immunization with plasmid DNA using a pneumatic gun. J Immunol Meth 1994;175:11-22.

22 Wagener S, Norley S, z. Megede J, Kurth R, Cichutek K: Induction of antibodies against SIV antigens after intramuscular nucleic acid inoculation using complex expression constructs. J Biotechnology 1996;44:59-65.

23 Wang R, Doolan DL, Le TP, Hedstrom RC, Coonan KM, Charoenvit Y, Jones TR, Hobart P, Margalith M, Ng J, Weiss WR, Sedegah M, de Taisne C, Norman JA, Hoffman SL: Induction of antigen-specific cytotoxic T lymphocytes in humans by a malaria DNA vaccine. Science 1998; 282:476-480.

24 Wolff JA, Lederberg J: A history of gene transfer and therapy. Gene therapeutics: methods and applications of direct gene transfer 1994:3-25.

25 Wolff JA, Malone RW, Williams P, Chong W, Ascadi G, Jani A, Felgner PL: Direct gene transfer into mouse muscle in vivo. Science 1990;247:1465-1468.

26 Yang NS, Sun WH: Gene gun and other non-viral approaches for cancer gene therapy. Nature Medicine 1995;1:481-483.

Dr. K. Cichutek, Department of Medical Biotechnology, Paul-Ehrlich-Institut,
Paul-Ehrlich-Str. 51-59, D-63225 Langen, Germany.

Brown F, Griffiths E, Horaud F, Schild GC (eds): A Celebration of 50 Years of Progress in Biological Standardization and Control at WHO
Dev Biol Stand. Basel, Karger, 1999, vol 100, pp 131-135

Control of Foot-and-Mouth Disease by Vaccination

F. Brown

Plum Island Animal Disease Center, Greenport, NY, USA

Foot-and-mouth disease is controlled in endemic areas by prophylactic vaccination. This paper describes the development of the methods that are used to produce these vaccines and the problems that still occur in the control of the disease. It also offers a glimpse of potentially new vaccines based on a molecular approach.

The year 1998 not only marks the 50th Anniversary of the World Health Organization, it also celebrates the 100th Anniversary of the publication by Loeffler and Frosch [1] of the paper describing their historic work which showed that foot-and-mouth disease (FMD) is caused by a virus. This is a landmark paper because FMD was the first animal disease shown to be caused by a filtrable agent.

The work of Loeffler and Frosch stemmed from the great need to control the disease because of the many outbreaks which were occurring in Germany at that time and the devastating effect it was having on the productivity of the infected animals. The clear goal was to control the disease by vaccination but it was not until 1925 that the first experimental vaccine was described [2]. The French group, led by Vallée and Carré, described a vaccine which they prepared by formaldehyde treatment of vesicular fluid collected from the lesions of an infected calf. This vaccine afforded protection against the disease but it was clear that vaccines prepared from virus grown in animals were unlikely to be sufficient for the large-scale production which would be necessary. Nevertheless, Rosenbusch et al [3] vaccinated more than one million cattle in Argentina in 1948 using what was essentially this method to prepare the vaccine.

It was the work of Frenkel [4] in Holland, however, that set the stage for the large-scale in vitro production of virus required for comprehensive vaccination programmes. Hecke [5] in Germany and the Maitlands [6] in England had grown the virus in vitro on a small scale in the early 1930s but, helped considerably by the contemporary discovery of antibiotics, Frenkel was able to grow the virus in fragments of bovine tongue epithelium in the large amounts required.

As tissue culture techniques were developed, it was soon possible to grow the virus in kidney cells from pigs and cattle. The most important development, however, was made by Mowat and Chapman in 1962 [7] when they showed that the virus grew to high titres in the baby hamster kidney cell line (BHK 21) described by MacPherson and Stoker [8]. This development was particularly important because

Capstick and his colleagues [9] were able to adapt this cell to grow in suspended culture, a considerable advantage for large-scale production.

One of the concerns at this stage in the development of FMD vaccines was their innocuity. Did the formaldehyde-inactivated vaccines still contain live virus? Moosbrugger [10] in 1948 was the first to express concern about their safety, a forerunner by six years of the problem which was encountered with the early Salk inactivated poliovaccine. In 1955, as a newcomer in virology, I was asked by Ian Galloway, the Director of the Animal Virus Research Institute at Pirbright, U.K., to look at this problem. From this study [11] it emerged that formaldehyde does not produce an innocuous vaccine, a conclusion reached at the same time by Graves [12]. Luckily, Weston Hurst, a scientist at Imperial Chemical Industries, Alderley Edge, Cheshire, U.K., had suggested to Galloway that N-acetylethyleneimine (AEI) should be tried. The success was immediate. Brown and Crick [13] showed that AEI inactivated FMDV by first order kinetics to produce highly immunogenic vaccines. A decade later, Bahnemann [14] showed that the imine itself, produced as required from bromoethylamine hydrobromide, is equally effective and overcomes the problems involved in storing AEI. This valuable innovation has been incorporated into FMD vaccine production methods worldwide.

Eradication of the disease from Western Europe by the application of these vaccines is a tribute to those involved in their development. But this goal could not have been achieved without the efficient application of the comprehensive annual vaccination programmes. The success in Western Europe and now in several countries in South America has shown that worldwide eradication is possible.

Antigenic Variation

A serious problem in the control of FMDV by vaccination had emerged in the early 1920s. Vallée and Carré [15] in France and Waldmann and Trautwein [16] in Germany showed that the virus occurred as three distinct serotypes, which came to be known as Vallée O, Vallée A and Waldmann C. This finding was very important because an animal which had been infected with virus of one serotype was still susceptible to infection by viruses belonging to the other two serotypes. To complicate the issue even more, antigenic variation within the individual serotypes was demonstrated by Bedson and his colleagues [17] in the late 1920s. The variation is often sufficient to render a vaccine ineffective against field outbreaks caused by a virus of the same serotype and it remains a problem even today.

As an added problem, the Pirbright group of Galloway, Brooksby and Henderson described four more serotypes in the 1950s [18]. These were found in Southern African Territories (SAT1, SAT2 and SAT3) and in Asia (Asia 1). Thankfully, no more serotypes have been found since 1954.

The problem of antigenic variation within a serotype has frequently been the cause of breakdown of vaccination programmes. This occurs when new field strains emerge which are only distantly related antigenically to those being used for vaccine production. For example, a significant outbreak occurred in Western Europe in the mid-1960s when an antigenic variant of serotype O broke through in the vaccinated population. It is clear that some of these outbreaks are caused by the introduction of virus in infected products. However, I consider that the emergence of these variants may also be related to the survival of the virus in the field, probably in a species which does not show clinical signs of the disease.

Molecular Approaches to Vaccination

The dominance of the empirical approach to the production of vaccines has been tempered since the mid-1950s by the entry into the field of personnel who were interested in molecules. It was soon clear from their work that only one virus-specific component in virus harvests, the 146S virus particle, was important in eliciting the neutralizing antibodies necessary for protection (Table 1). Moreover, the immunogenicity of the particle was reduced by more than 100-fold by lowering the pH below 7 [19], and with some strains of the virus a similar loss was incurred by incubating them with trypsin [20]. The step required to explain these observations was to purify the virus, preferably labelled with precursors of RNA and proteins. This was achieved in 1963 [21] and the subsequent analysis showed that the virus, like poliovirus, consisted of one molecule of RNA encapsulated in a 30nm spherical particle consisting of 60 copies of each of four proteins (VP1-VP4) and, interestingly, one copy of the virus RNA polymerase [22]. This enzyme can also function as a nuclease [23] and is responsible for the loss of infectivity when the virus is kept at room temperature.

Lowering the pH below 7 causes the virus particle to explode, releasing the infectious RNA and yielding a 12S particle consisting of a pentamer of VP1, VP2 and VP3 and an aggregate of VP4. The loss of infectivity and immunogenicity when the virus particle was treated with trypsin was explained by the cleavage of VP1. Otherwise the particles appeared to be unchanged, as judged by electron microscopy, but they no longer reacted with IgM antibody.

Once nucleic acid sequencing became available it was possible to identify the cleavage sites on VP1 [24]. One sequence of the protein (at amino acids 138-154) and a second sequence (202-213 at the C terminus) were removed by the enzyme. These observations provided clues about the immunogenic epitopes of the virus particle, and subsequent synthesis of peptides corresponding to the sequences showed that both elicited neutralizing antibodies in guinea pigs [25]. The sequence encompassing amino acids 141-160 had considerably higher immunogenic activity and, significantly, contained the RGD triplet which is involved in cell attachment and entry of the virus into cells. The RGD sequence is present in all the FMD virus VP1 sequences which have been analysed.

These results indicated that a totally synthetic vaccine, not involving infectious virus, was within the realms of possibility. Subsequent work has shown that, provided the T-cell epitopes relevant to the species being vaccinated are included in the synthetic vaccine, protection against challenge infection can be achieved.

Table 1: Antigenic components of foot-and-mouth disease virus.

Component	**Sedimentation constant**	**Composition**
Virus particle	146S	1 molecule ssRNA ($M_r = 2.6 \times 10^6$), 60 copies of each of VP1-3 (M_r c. 24×10^5) and VP4 (M_r c. 8×10^5)
Empty particle	75S	60 copies of VP0[3], VP1 and VP3
Protein subunit	12S	Pentamer of VP1-3
Virus infection associated antigen	3.8S	RNA polymerase (M_r c. 56×10^3)

* VPO comprises VP4 and VP2 covalently linked.

It is of considerable significance that the region of VP1 which elicits the protective levels of antibodies varies greatly in amino acid sequence between serotypes. X-ray crystallography has shown that the sequence is present on a prominent loop region on the surface of the virus particle [26] and plays a key role in its attachment to susceptible cells.

The production of infectious cDNA molecules corresponding to the full-length RNA genomes of two of the serotypes of the virus [27, 28] has enabled the sequence to be manipulated to produce mutants possessing biological properties differing from those of the «parent» viruses. Work in this area has so far been restricted to the major immunogenic epitope of the virus but the technology offers great opportunities for providing information regarding such properties as pathogenicity and host range.

It is probably too early even to hazard a guess at the eventual outcome of vaccines based on the direct application of the DNA coding for the epitope of possible significance in eliciting the appropriate immune responses. Experiments designed to compare the immune responses to DNA vaccines and the proteins which they encode have escaped my attention. It is almost certain, however, that vaccines in the next 50 years will be based on the molecular advances which have been made during the past 50 years.

References

1 Loeffler F, Frosch P: Berichte der Kommission zur Erforschung der Maul-und-Klauenseuche bei dem Institut fuer Infektionskrankheiten in Berlin. Centralblatt Bakt I Abt Orig 1898;23: 371-391.

2 Vallée H, Carré H, Rinjard R: On immunisation against foot-and-mouth disease. Rech Med Vet 1925;101:297-299.

3 Rosenbusch CT, Decamps A, Gelormini N: Intradermal foot-and-mouth disease vaccine. Results obtained from the first million head of cattle vaccinated. J Am Vet Med Assn 1948;112: 45-47.

4 Frenkel HS: La culture du virus de la fièvre aphteuse sur l'épithelium de la langue des bovidés. Bull Off Int Epizool 1930;28:155-162.

5 Hecke F: Zuechtungsversuche des Maul-und-Klauenseuchevirus in Gewebekulturen. Zentbl Bakt Parasitkde 1931;I:386-414.

6 Maitland MC, Maitland HB: Cultivation of foot-and-mouth disease virus. J Comp Path Therap 1931;44:106-113.

7 Mowat GN, Chapman WG: Growth of foot-and-mouth disease virus in a fibroblastic cell line derived from hamster kidneys. Nature 1962;194:253-255.

8 MacPherson I, Stoker M: Polyoma transformation of hamster cell lines – an investigation of genetic factors affecting cell competence. Virology 1962;16:147-151.

9 Capstick PB, Telling RC, Chapman WG, Stewart DL: Growth of a cloned strain of hamster kidney cells in suspended cultures and their susceptibility to the virus of foot-and-mouth disease. Nature 1962;195:1163-1164.

10 Moosbrugger GA: Recherches expérimentales sur la fièvre aphteuse. Schweiz Arch Tierheilk 1948;90:179-198.

11 Brown F, Hyslop NStG, Crick J, Morrow AW: The use of acetylethyleneimine in the production of inactivated foot-and-mouth vaccines. J Hyg 1963;61:337-344.

12 Graves JH: Formaldehyde inactivation of foot-and-mouth disease virus as applied to vaccine preparation. Am J Vet Res 1963;24:1131-1135.

13 Brown F, Crick J: Application of agar gel diffusion analysis to a study of the antigenic structure of inactivated vaccines prepared from the virus of foot-and-mouth disease. J Immunol 1958; 82:444-447.

14 Bahnemann HG: Binary ethyleneimine as an inactivant for foot-and-mouth disease virus and its application for vaccine production. Arch Virol 1975;47:47-56.

15 Vallée H, Carré H: Sur la pluralité du virus aphteux. Cptes rendus hebd Acad Sci 1922;174:1498-1500.

16 Waldmann O, Trautwein K: Experimentelle Untersuchungen uber die Pluralitat des Maul-und-Klauenseuche Virus. Berlin tierarztl. Wschr 1926;42:569-571.

17 Bedson SP, Maitland HB, Burbury YM: 2nd Report, Foot-and-Mouth Disease Research Committee. HMSO London 1927;99.

18 Brooksby JB: The virus of foot-and-mouth disease, in Smith KM (ed): Adv in Vir Res. Academic Press, 1958, vol 5, pp 1-37.

19 Brown F: Picornaviruses, in van Regenmortel MHV, Neurath AR (eds): Immunochemistry of Viruses. The Basis for Serodiagnosis and Vaccines. North Holland, Elsevier, 1990, vol 2, pp 153-169.

20 Wild TF, Burroughs JN, Brown F: Surface structure of foot-and-mouth disease virus. J Gen Virol 1969;4:313-320.

21 Brown F, Cartwright B: Purification of radio-active foot-and-mouth disease virus. Nature 1963; 31:1168-1170.

22 Newman JFE, Piatti PG, Gorman BM, Burrage TG, Ryan MD, Flint M, Brown F: Foot-and-mouth disease virus particles contain replicase protein 3D. Proc Nat Acad Sci 1994; 91:733-737.

23 Denoya CD, Scodeller EA, Vasquez C, La Torre JL: Foot-and-mouth disease virus: II Endoribonuclease activity within purified virions. Virology 1978;89:67-74.

24 Strohmaier K, Franze R, Adam K-H: Localisation and characterisation of the antigenic portion of the foot-and-mouth disease virus protein. J Gen Virol 1982;59:295-306.

25 Bittle JL, Houghten RA, Alexander H, Shinnick TM, Sutcliffe JG, Lerner RA, Rowlands DJ, Brown F: Protection against foot-and-mouth disease by immunization with a chemically synthesized peptide predicted from the viral nucleotide sequence.Nature 1982;298:30-33.

26 Acharya R, Fry E, Stuart D, Fox G, Rowlands DJ, Brown F: The three-dimensional structure of foot-and-mouth disease virus at 2.9 Å resolution. Nature 1989;337:709-716.

27 Zibeck A, Maass G, Strebel K, Falk MM, Beck E: Infectious foot-and-mouth disease virus derived from a cloned full-length cDNA. J Virol 1990;64:2467-2473.

28 Rieder E, Bunch T, Brown F, Mason PW: Genetically engineered foot-and-mouth disease viruses with poly (C) tracts of two nucleotides are virulent in mice. J Virol 1993;67:5139-5145.

Prof. F. Brown, Plum Island Animal Disease Center, P.O. Box 848, Greenport, NY 11944, USA.

Brown F, Griffiths E, Horaud F, Schild GC (eds): A Celebration of 50 Years of Progress in Biological Standardization and Control at WHO.
Dev Biol Stand. Basel, Karger, 1999, vol 100, pp 137-141

Standardisation and Licensing of Combination Vaccines Based on DTP

I. Gust

Research and Development Director, CSL Limited, Melbourne, Australia

Abstract: Development of combination vaccines is a time consuming and expensive exercise with significant technical, clinical, regulatory, manufacturing and marketing challenges, which is likely to result in a rationalisation of production worldwide.

INTRODUCTION

For those of us who have grown up in an era of widely available vaccines and antibodies, it is difficult to visualise how the world was 50 or 60 years ago when infectious disease were a major cause of morbidity and mortality, especially among young children. One has only to walk through a churchyard and look at the headstones to recognise how common it was for families to lose a child early in life – a situation which is now largely restricted to the poorest countries in the developing world.

From the 1920s, a range of vaccines and antisera were developed which had a profound effect on the incidence of some of the more common childhood diseases and their sequelae. The first products to be made were relatively crude and based on simple, widely available technology, first toxoids to prevent diphtheria and tetanus, then inactivated whole cell vaccines to prevent pertussis and in the decade after World War II, vaccines to prevent influenza and poliomyelitis. Initially, each of these was monovalent, but in the 1940s for a combination of convenience, compliance, coverage and cost, the idea of combining vaccines arose – initially diphtheria and tetanus toxoids to give DT, later diphtheria and tetanus toxoids and whole cell pertussis vaccines to provide DTP. This combined vaccine has provided the foundation of childhood immunisation programmes for the past 50 years and its wide use has led to a greater than 98% reduction in the prevalence of these diseases in the developed world, and through the efforts of the Expanded Program of Immunisation, an approximately 80% reduction worldwide.

While DT and DTP were the first combination vaccines, they are not unusual. Some of our most widely used vaccines, such as those used to prevent influenza, poliomyelitis and measles, mumps and rubella are trivalent, whilst the currently used pneumococcal vaccine is 23 valent. Interestingly some of the issues we are grappling with today were widely debated in the 1940s and 1950s.

History of the Vaccine Industry

For much of this century, the history of the vaccine industry has largely been the history of DTP production. From World War I until the 1970s, most of the vaccines in the world were produced in government funded, government run facilities, using simple, widely available technology. As regulatory requirements were modest, development times were short and large volumes of product could be made at relatively low cost without the requirement for precious foreign exchange. Over the past 20 years, a major revolution has taken place in the developed world; increasingly complex production methods and increasingly stringent regulatory requirements have increased the cost of developing new vaccines, added to the lead times, and inevitably to the cost of the final vaccines. As governments have been unable to provide the resources to fund these activities, they have left the field, so that today more than 90% of the vaccines used in the developed world are produced in the private sector by four major manufactures.

The benefits of long-term government investment in basic research and industry's ability to convert innovation into new products has been demonstrated by the development and registration of several new vaccines – hepatitis B, haemophilus influenza b, hepatitis A, several acellular pertussis vaccines and human rotavirus vaccine, which have their maximum benefit when administered within the first year of life.

As a fully immunised child may require multiple contacts and around a dozen injections in the first year of life, it seemed logical to attempt to combine them into larger (quadra-, penta- and hexa-valent) combinations. Most of the newer combinations are based on DTP as this vaccine is generally administered three times in the first year of life with a booster (or boosters) later as the diseases against which it is aimed are unlikely to be eradicated.

Development of Modern Combination Vaccines

While it may seem relatively straightforward to combine existing registered vaccines such as DTP and HB and/or Hib, let me assure you that this is not the case. The most experienced companies in the world are taking between five and 10 years to develop combination vaccines, at a cost of many tens of millions of dollars, and some combinations are proving extremely difficult to formulate. Vaccine combination thus provides a formidable array of technical, clinical, regulatory, manufacturing and marketing challenges.

Technical

While many components of combination vaccines are adjuvanted with alum, the adjuvants used by different manufacturers differ not only in their basic composition (hydroxide or phosphate) but in their adsorptive properties. A significant amount of reformulation may be needed to obtain a product with suitable properties.

Addition of multiple components to a vaccine increases the risk that interference will occur or that the resultant product will develop new and undesirable properties. Our experience at CSL in formulating a DTPw-HB-Hib vaccine is illustrative. When we initially mixed the three licenced vaccines, each of which was formulated as a stable solution, clumping occurred. It was necessary to reformulate our DTP and evaluate more than 20 options before selecting a product which would accept other antigens, was readily resuspended, and had an appropriate shelf life.

Assays provide an additional challenge, as antigens when combined may behave differently than when presented alone. We have found that an EIA which works well for monovalent Hib was unsatisfactory when that vaccine was combined with DTP-HB and we had to establish and validate an alternative HPLC assay. Another aspect which is often overlooked is the difficulty in transferring assays from the site of production to the site of formulation and establishing concordance. Even with high quality staff obtaining concordance in assays performed in different continents with different reagents and equipment is challenging.

Clinical challenges

What has been learned is that no two combination vaccines are the same; each new combination needs to be evaluated in the clinic for reactogenicity and immunogenicity (by a variety of schedules) and its capacity, or otherwise, to interfere with other vaccines given at the same time. In general each combination is compared with its component vaccines given separately. Given that it is no longer possible to evaluate clinical efficacy directly, surrogate markers such as antibody levels and protection in an animal model are required.

While there are soundly based serological correlates of protection for D, T, HB and Hib, there are no generally agreed correlates of protection against pertussis. The mouse intracerebral challenge test (Kendrick assay) has proved to be a useful potency and release assay for whole cell pertussis vaccines and is assumed to be equally valuable for whole cell based combination vaccines. Unfortunately this assay is unsuitable for acellular vaccines.

Given the difficulty of gathering clinical data on the efficacy of acellular based combination vaccines, an animal model which is capable of measuring potency is urgently needed. The recently described intranasal and aerosol challenge models, once standardised and validated may solve this dilemma and the forthcoming WHO Consultation at NIBSC on the subject is an important step.

One area that has had little attention is interchangeability of products. While this is of interest to regulatory authorities and public health officials, there is little incentive for manufacturers to undertake expensive and time-consuming studies of all of the possible permutations. While common sense dictates that it is preferable to complete a schedule with the same vaccine, there are no current data to suggest that beginning with a product from one licenced manufacturer and concluding with another affects the outcome.

Regulatory challenges

The problem for regulators is that each combination needs to be considered separately as the components and processes are product specific. It should be possible to simplify the review process by focussing on those aspects where the process varies from that of the registered components and establishing consistency of manufacture.

The critical regulatory issue is whether combination vaccines should induce equivalent serological responses to their separate components or whether the responses should be clinically adequate? This is an important issue as most pentavalent vaccines developed to date have resulted in a reduced immune response to one component (usually Hib but sometimes HB). The question is whether this is clinically important? Agencies will need to exercise common sense.

Finally the issue of appropriate product information needs to be addressed; as combinations extend to five, six, or seven components, the amount of information that could be included in a PI has become voluminous. For practical reasons, the PI should be restricted to critical information.

Manufacturing challenges

There are a number of formidable manufacturing challenges as most combination vaccines require transport of one or more components from a distant site. Careful control and documentation is required for each stage in the process and evidence that the final product can be made consistently is particularly important.

Marketing challenge

As the use of combination vaccines is market driven, manufacturers need to ensure that their combinations and schedules are appropriate for each market. This may require multiple clinical trials, using different schedules in different countries and the ability to respond rapidly to changes in government policy.

The Future

Currently about 60% of the world's DTP is produced by local manufacturers and a significant quantity is regarded by WHO as not of known good quality. Given that the ability to sustain a vaccine industry is partly dependent on the ability of manufacturers to access and develop new products and that these are often complex to produce and are protected by patents, the decline in local manufacture, which has been a feature of the past two decades, seems certain to continue. Local manufacturers face a number of options from waiting until the products are no longer under patent, to local bottling and filling, and to joint ventures. In my opinion, what we are witnessing is a form of economic Darwinism in which only institutions with high quality facilities, staff and management, a pathway to accessing new technology and intellectual property (IP), substantial financial resources and access to major markets will thrive

The Role of WHO

What should WHO's role be in this rapidly changing scene? To my mind, the organisation's main responsibility is to ensure the quality of vaccines used in international control programmes. As such, it has a vital role to play in establishing guidelines for production and control, developing standards and reference preparations, assisting National Control Authorities to improve their skills and promoting an environment in which combination vaccines will be widely used in developing countries which will then be able to benefit from the economies of large scale production.

Dr. I. Gust, Research and Development, CSL Limited, 45 Poplar Road, Parkville, Victoria 3052, Australia.

Brown F, Griffiths E, Horaud F, Schild GC (eds): A Celebration of 50 Years of Progress in Biological Standardization and Control at WHO.
Dev Biol Stand. Basel, Karger, 1999, vol 100, p 143

SESSION IV

International Perspectives

Co-chairs: *Geoffrey C. Schild (U.K.)*
and Ian Gust (Australia)

Brown F, Griffiths E, Horaud F, Petricciani GC, Schild JC (eds): A Celebration of 50 Years of Progress in Biological Standardization and Control at WHO.
Dev Biol Stand. Basel, Karger, 1999, vol 100, pp 145-148

Regulation of Biologicals in the European Union

G. Vicari

Roma, Italy

INTRODUCTION

Biological and biotechnology-derived pharmaceuticals represent a peculiar class of medicinal products, being complex, high molecular weight molecules, which often cannot be completely defined by conventional physical and chemical tests, and which may require immunological, biochemical or bioassay techniques to measure the quantity present and to assess their activity. Such characteristics raised the concern of the regulatory authorities on the safety and efficacy of these products since the introduction of the first biologicals and national regulation was introduced already in the 1920s-1930s in several European countries. Moreover, a special office was established at the international level under the League of Nations, taking care of the programme of biological standardization, at present carried out by the Biologicals Unit of the World Health Organization.

Since the 1980s an evolution from classic biologicals to biotechnology-derived products has occurred as a consequence of the scientific progress, which has influenced manufacturing, purification and analytical procedures. The range of biological products has increased tremendously, including many biotechnologically-derived products such as insulin, hormones, interferons, erythropoietins, growth factors, cytokines and monoclonal antibodies.

EUROPEAN RULES GOVERNING HIGH TECHNOLOGY MEDICINAL PRODUCTS

In the interest of public health and to promote the free movement of medicinal products in the European Union, legislative measures were adopted to harmonize the conditions of manufacture, testing and authorization of medicinal products in the Member States. In this context, the European Commission submitted to Council in September 1984 a package of regulatory measures in favour of high technology medicinal products, particularly those produced by biotechnology. These measures, adopted by the Council in December 1986, came into force in July 1987.

For biotechnology-derived medicinal products, the Directive 87/22/EEC, at present included in the Council Regulation 2309/93/EEC, obliged the competent authorities to consult each other systematically within the Committee for Proprietary Medicinal Products (CPMP) before taking any decision on recombinant DNA-technology, controlled expression of genes and hybridoma/monoclonal antibody methods. In accordance with Directive 87/22/EEC, the Council adopted in 1989 two Directives (89/342/EEC and 89/381/EEC) to extend the scope of pharmaceutical legislation to cover immunological products and plasma-derived products respectively. In January 1995 a new European system for the authorization of medicinal products came into effect. The European Agency for the Evaluation of Medicinal Products (EMEA) was established by the European Council and London was chosen as its seat. [1, 2]

Biotechnology Working Party

To assist the Commission and the competent authorities, a special EEC Biotechnology/Pharmacy Working Party [3] was set up in 1986 with two main tasks: (i) to advise the CPMP on individual applications for marketing authorization for medical products derived from biotechnology submitted in accordance with the EEC centralized procedure; (ii) to establish specific guidelines on the quality and safety of these products.

At present, the Group is denominated Biotechnology Working Party (BWP) and it continues to fulfill its tasks. Moreover, it cooperates with the CPMP in the programme of harmonization established under the International Conference for Harmonization (ICH) directed to establish harmonization between EU, Japan and USA.

Guidelines

During the past twelve years the BWP has put considerable effort into the preparation of Guidelines on data requirements for biological and biotechnology-derived medicinal products. Major guidelines have been prepared on the following issues:

- production and quality control of medicinal products derived by recombinant DNA technology;
- quality of biotechnological products: analysis of the expression construct in cell lines used for production of r-DNA derived proteins (ICH guideline);
- quality of biotechnological products: stability testing of biotechnological/biological products (ICH guideline);
- quality of biotechnological products. Derivation and characterization of cell substrates used for the production of biotechnological/biological products (ICH guideline);
- production and quality control of cytokine products derived from biotechnology;
- gene therapy product quality aspects in the production of vectors and genetically modified somatic cells;
- use of transgenic animals in the manufacture of biological medicinal products for human use;

- production and quality control of monoclonal antibodies;
- note for guidance on plasma-derived medicinal products;
- note for guidance on harmonization of requirements for influenza vaccines;
- note for guidance on pharmaceuticals and biological aspects of combined vaccines;
- allergen products.

Important safety issues have also been covered by the following guidelines:

- virus validation studies: the design, contribution and interpretation of studies validating the inactivation and removal of viruses;
- validation of virus removal/inactivation procedures: choice of viruses;
- minimizing the risk of transmission of agents causing spongiform encephalopaties via medicinal products;
- viral safety of medicinal products derived from cell cultures (ICH guideline);
- safety evaluation of Biotechnology Products (ICH guideline).

The BWP will continue its activities on biological and biotechnology-derived medicinal products according to the mandate given by the CPMP. Priorities in the immediate future are: (i) revision of the gene therapy guideline; (ii) DNA vaccines; (iii) CJD and plasma-derived products, and (iv) viral safety.

Future Perspectives and Final Considerations

All these activities, past, present and future, require and will require a strict liaison with WHO, European Pharmacopoeia and European Commission, as well with the various manufacturers associations. Hopefully, the process of harmonization in the framework of the ICH will also continue.

Problems exist and new ones may arise in future developments and use of biologicals and a fruitful collaboration between the above indicated organizations is necessary in order to solve such problems and attain international harmonization. As an example, the ICH has not taken into account the plasma-derived medicinal products and many differences still exist between the EU and the US, such as the HIV p24 antigen testing, required in the US and not considered in Europe, the nucleic acid amplification test (NAT) for the detection of HCV RNA, performed in plasma pools in Europe and in some final products in the US.

Relationship with WHO poses more difficult problems, which go beyond the scientific levels and need attention at the political and financial level. WHO activity is mainly directed to the developing world. Its scientific approach does not differ from the approach of the EMEA or of the FDA in evaluating the quality, efficacy and safety of biologicals. However, new biologicals coming to our attention are more sophisticated and more expensive, which the developing world cannot afford. A typical example is represented by the combined vaccines, whose development was stimulated by WHO to allow the simultaneous immunization of children of the developing world against several agents. Purchase of such vaccines cannot be afforded by the countries of the developing world, while they have been introduced in the advanced countries and used mainly on the basis of a better compliance for the fortunate children of EU and US. Dr. G. H. Brundtland, new Director General of WHO, has recently called upon governments to acknowledge their responsibility for health. In her address to the Health Assembly she envisaged «a world where solidarity binds the fortunate with those less favoured: Where our collective efforts will help to roll back all the diseases of the poor. Where our collective efforts assure universal access to compassionate and competent health care.» [4] Our efforts should be combined to attain such a world.

References

1 Brunko P, Sauer F: EEC legislation of medicinal products of biological origin, in Brown F, Horaud F (eds): Virological Aspects of the Safety of Biological Products. Dev Biol Stand. Basel, Karger, 1991, vol 75, pp 21-25.

2 European Commission. The rules governing medicinal products (EUDRALEX). European Communities, Luxembourg, 1998.

3 Schild GC: EC ad hoc Working Party on Biotechnology/Pharmacy. Summary of activities 1986-1990, in Brown F, Horaud F (eds): Virological Aspects of the Safety of Biological Products. Dev Biol Stand. Basel, Karger, 1991, vol 75, pp 247-250.

4 WHO: Notes & News. New Director General of WHO. World Health Forum 1998;19:332-333.

Dr. G. Vicari, V. Principessa Clotilde 5, I-00196 Roma, Italy.

Brown F, Griffiths E, Horaud F, Schild GC (eds): A Celebration of 50 Years of Progress in Biological Standardization and Control at WHO.
Dev Biol Stand. Basel, Karger, 1999, vol 100, p 149

Assuring the Safety of Plasma Products: a Developing Country's Perspective

D. Armstrong

Natal Bioproducts Institute, Pine Town, South Africa

Abstract: There is interdependence between blood transfusion technology and fractionation. Assuring the safety of plasma products is largely dependent on the establishment of an effective blood transfusion service, as this in turn will maximise the safety of the plasma supply. This is outlined in the World Health Organisation (WHO) Aid Memoir on Blood Safety, which includes a component therapy programme and the implementation of the required safety support systems.
In a developing country like South Africa, affordability, sustainability, and accessibility linked to self sufficiency are important criteria to be considered when deciding on a product profile, the technology to use and whether or not to establish a fractionation facility. In collaboration with health care providers, cognisance has to be taken of the WHO and the country's specific Essential Drug Lists, as well as the needs of the user and patient groups.
Safety, quality and efficacy cannot be compromised. However, a key word in this regard is «appropriate», particularly when selecting technology and type of product. It is better to have an adequate supply, of a higher yielding, single viral-inactivated, intermediate purity factor VIII than insufficient, unaffordable, double viral-inactivated, high purity product.
Worldwide, in both the «for profit» and «not for profit» sectors, rationalisation and centralisation of fractionation facilities have increased the minimum volumes of plasma that will be considered for processing. This can preclude smaller, developing countries from participation in contract fractionation. There is a need for regional co-operation and the influential stakeholders would be the WHO and the European Plasma Fractionation Association (EPFA).

Dr. D. Armstrong, Natal Bioproducts Institute, 10 Eden Road, Private Bag X9043, Pine Town 3600, South Africa.

Brown F, Griffiths E, Horaud F, Schild GC (eds): A Celebration of 50 Years of Progress in Biological Standardization and Control at WHO.
Dev Biol Stand. Basel, Karger, 1999, vol 100, pp 151-153

ICH Activities on Biotech Topics

P. Zorzi-Morre

Sanofi Recherches, Labège, France

The ICH expert working group involved in the quality control of biotech products was set up in 1992 and results were presented at three international conferences in 1993, 1995 and 1997. Five topics have been harmonized since then: (i) viral safety; (ii) genetic stability; (iii) product stability; (iv) cell substrates; and (v) product specifications. From the very first discussion in 1992, it was apparent that biotechnology was a complex area which could not be defined spcifically and consequently the scope was defined separately for each topic.

The guideline «viral safety» [1] is concerned with testing and evaluating the viral safety of biotechnology products derived from characterized cell lines of human or animal origin. The importance of this topic was stressed by the three regions (Europe, USA, Japan) even if biotechnology products derived from cell lines have not yet been implicated in the transmission of viruses. Three principal, complementary approaches have evolved to control potential viral contamination: (i) testing of cell lines and raw materials for absence of viruses; (ii) assessing the capacity of the production process to clear the potential presence of infectious viruses; and (iii) testing the product for absence of viruses. The cell line must be qualified by appropriate tests for detection of retroviruses, endogenous and non-endogenous viruses. Tests are performed at various levels, including Master Cell Bank, Working Cell Bank and cells at the limit of in vitro age used for production. Appropriate testing for viruses is applied to the unprocessed bulk which represents one of the most suitable levels at which the possibility of adventitious virus contamination can be determined with a high probability of detection. Data must be provided on three batches as well as continuing assessment on production batches. According to the results of virus tests obtained at the cell line level and at the unprocessed bulk level, the guideline also provides an action plan to be applied for further viral testing in term of viral clearance studies and tests on the purified bulk. Five cases are considered, the most common situations being case A (no virus detected) and case B (rodent retroviruses). Cases C, D and E for which there is presence of a virus other than a rodent retrovirus should be discussed with the regulatory authorities.

The «Genetic Stability» [2] document presents guidance regarding the characterization of the expression construct for the production of recombinant DNA protein products in eukaryotic and prokaryotic cells. The rationale for the evaluation of the integrity of the expression construct is to demonstrate that the correct coding sequence has been incorporated and is maintained to the end of production; This evaluation can be performed either by nucleic acid analysis or protein analysis. The relative importance of both analyses will vary from product to product.

The «Biotech Product Stability» [3] document is an annex to the guideline on chemical entities [4] and covers the well-characterized proteins and polypeptides, their derivatives and products of which they are components, and which are isolated from tissues, body fluids, cell cultures, or produced using rDNA technology. Even if it is obvious that biotech products have some specificities which are highlighted in the guideline, the major principles laid down in the main guideline applies also to biotech products: (i) stability data at submission must be given on three batches; (ii) pilot scale is permitted with a commitment to place the first three manufacturing batches into the long-term stability programme after approval; (iii) the quality of the stability batches should be representative of the quality of the material used in preclinical and clinical studies and the quality the material to be made at manufacturing scale; (iv) the expiry date should be based on real-time/real-temperature data.

The objective of the guideline on «Cell Substrates» [5] is to provide guidance on the derivation and characterization of human and animal cell lines and microbial cells to be used to prepare biotechnological/biological products, and for the preparation and characterization of cell banks to be used for production. Documentation on source, history and generation of the cell substrate need to be provided to facilitate an overall evaluation which will ensure the quality and safety of the product. Manufacturers should describe their cell banking system, the most practical approach being a two-tiered system in which the Master Cell Bank (MCB) is used to generate Working Cell Banks (WCBs). The characterization and testing of banked cell substrates confirm the identity, purity, and suitability of the cell substrates for manufacturing use. Manufacturers should perform tests for identity and purity once for each MCB, and tests for stability during cell cultivation once for each product to be registered. In addition, tests of purity and limited tests of identity should be performed once on each WCB.

«Specifications for Biotech Product» [6] was identified as a fundamental topic from the beginning of the discussion in 1992, but it was felt that priority should be given to the four other topics, thereby establishing the basis for product specifications. The guideline gives a definition for specification which is common with that for chemicals [7] and covers proteins and polypeptides which can be highly purified and characterized. It provides the rationale for setting specifications and lists the items to be addressed for the release of drug substance and drug product. Other points of harmonization reached are the concept of heterogeneity, well acknowledged for this class of product, biological activity, purity and impurities, reference standards/materials.

International harmonization of these five topics provides a solid foundation for the «Common Technical Document» (CTD/M4) for biotech products which is presently discussed in ICH. This document should facilitate harmonization of the criteria used to evaluate comparability of products in the case of manufacturing changes and multi-use facilities, an important issue for the pharmaceutical industry.

References

1 Viral safety evaluation of biotechnology products derived from cell lines of human or animal origin. (Q5A)

2 Analysis of the expression construct in cells used for production of rDNA derived protein products. (Q5B)
3 Stability testing of biotechnological/biological products. (Q5C)
4 Stability testing for new drug substances and products. (Q1A)
5 Derivation and characterization of cell substrates used for production of biotechnological/biological products. (Q5D)
6 Specifications : tests procedures and acceptance criteria for biotechnological/biological products. (Q6B)
7 Tests procedures and acceptance criteria for new drugs substances and products (Q6A)

Dr. P. Zorzi-Morre, Sanofi Recherche, Centre de Labège, Bureau N-218, B.P. 137,
F-31676 Labège Cedex, France.

Brown F, Griffiths E, Horaud F, Schild GC (eds): A Celebration of 50 Years of Progress in Biological Standardization and Control at WHO.
Dev Biol Stand. Basel, Karger, 1999, vol 100, pp 155-158

Regulation of Biologicals in China

H. Zhou

National Institute for the Control of Pharmaceutical & Biological Products, Beijing, China

Biological products in China are regarded as the main weapon for prevention, treatment and diagnosis of infectious diseases as well as other diseases. Since the founding of the People's Republic of China, «Putting Prevention First» has been raised as one of the principles of health care for prophylaxis and treatment of diseases. Therefore, great attention has been paid to the production and quality control of biological products.

Not long after the founding of new China, six institutes for research and production of biological products, namely Beijing, Changchun, Chengdu, Lanzhou, Shanghai and Wuhan Institutes for Biological Products, were set up respectively in the six administrative regions of China, thus forming a national network for production, supply and immunization of biological products. Until now, more than 200 biologicals have been produced in China. Among them there are 38 belonging to major biological products ranging from 13 viral vaccines, 10 bacterial vaccines, eight blood products, five cytokines and two antitoxins and toxoids. The domestic demand for those main products used for the purpose of prevention and treatment can be met.

China joined WHO in the Expanded Programme on Immunization in 1981 and promulgated the Regulations of National Immunization Programme in 1982. In response to the call of the government, a nationwide campaign was launched in the immunization with DPT, BCG, measles and polio vaccines in China. In addition, no cases of disease have been found caused by wild poliovirus strains. It is predicted that poliomyelitis will have been eradicated by the end of this century.

As biological products are derived from biological sources, the quality of the products is of critical importance. To ensure the quality of biological products, the Biological Division of Drug Administration Bureau of the Ministry of Health exercises the administrative functions. In addition, the National Institute for the Control of Pharmaceutical and Biological Products (NICPBP) also takes responsibilities on behalf of the state for the supervision of product quality. Recently the new agency of State Drug Administration (SDA) has been formed directly under the leadership of the State Council. It comprises four departments dealing with professional administrative supervision. For example, the Department of Drug Regulation has the functions of drug registration and specification in which there is a division specifically to manage the biological products. The Department of Safety Monitoring and Inspection has the functions for the control of preclinical data and inspection of facilities including GLP, GCP and GMP compliance. The Department of Market Compliance has the function of drug supervision after a drug has been put onto the market. The Department of Medical Devices has the function of registration and supervision for medical devices. The new SDA comprises the whole regulatory functions of drug and medical devices.

With regard to drug regulation, as early as 1985, the Drug Administration Law of the People's Republic of China was put into effect. In the same year, Provisions for the Evaluation and Approval of New Biologicals was implemented (Table 1). According to the provisions, new biologicals should be classified into four categories:

Class 1: attenuated and unattenuated live vaccines;

Class 2: dead vaccines, toxoid, antitoxins, specific immunoglobulins, antisera, mycoplasma, in vivo diagnostic reagents;

Class 3: human plasma derivatives, immunological preparations derived from tissue and plasma with human & animal origin;

Class 4: in vitro diagnostic reagents for immunological & serological purposes.

Each category requires specific technical data for application. To the end of 1997, 238 kinds of new biologicals were approved.

Recombinant DNA and hybridoma technologies were introduced to China in 1975 and 1979. They were used in research and development in diagnosis and therapy. The development of biotechnology has had a great influence on various fields of national economy. The Ministry of Public Health has developed «Points to consider for assuring the quality of the products prepared by recombinant DNA technology» and «Points to consider for assuring the quality of monoclonal antibodies of murine origin for human use» in 1990. Since then, rapid progress has been made in research and development of biotechnology. Seven products such as Interferon α and γ, hepatitis B, interleukin 2, basic fibroblast growth factor GM-CSF erythropoietin etc. have been put in production; 10 of its kind have been at the stage of clinical trial; about six of them have been given IND approval (Table 2).

Table 1: Biological regulatory directives.

• Drug Administration Law of P.R.C.	1985
• Provisions for the Evaluation of New Biologicals	1985
• Points to consider for assuring the quality of the products prepared by recombinant DNA technology	1990
• Points to consider for assuring the quality of monoclonal antibodies of murine origin for human use	1990
• National Committee on Biological Standardization	1988
• Good manufacturing practice	1988
• Provisions for the control of biologicals	1993
• Provisions for the control of blood products	1996

Table 2: Status of r-DNA products in China (1997).

Name	Status
Interferon α 2b	production
Interferon α 1b	production
Interferon α 2a	production
Interferon γ	production
EPO	production
Hepatitis B (CHO)	production
Hepatitis B (yeast)	production
G-CSF	production
Interleukin-2	production
b-FGF	production
Growth hormone	clinical trial
GM – CSF	clinical trial
Interleukin-2/125 ser	clinical trial
Interleukin-2/125 ala	clinical trial
Interleukin-3	clinical trial
Streptokinase	clinical trial
Staphylokinase	IND
TNF α	IND
TNF α derivative 3a	IND
EGF	IND

In 1988 the National Committee on Biological Standardization was set up, the predecessor of which was called the «Biological Committee», founded in the 1950s. The main function of the Committee is to work out the China Requirements for production and quality control of biological products. 166 kinds of biologicals used for prevention, treatment and diagnosis were compiled in the current China Requirements of Biological Products.

In 1988, Good Manufacturing Practice for medicines in China was promulgated. In line with the WHO GMP revised in 1992, China GMP was supplemented by adding new regulations specifically for biological products. Quality control and batch release for biological products are conducted by the manufacturers' QC with regular sampling by the National Control Laboratory. NICPBP is responsible for the supervision of biological manufacturers in the whole country by taking samples regularly for testing and sending national drug inspectors to examine the GMP implementation of manufacturers.

Based on the experience over the past 20 years, the MOH worked out the Provision on Control of Biological in July 1993. In strengthening the control of blood products, the State Council specifically promulgated «Provision on Control of Blood Products» in December 1996.

To improve the quality of diagnostic reagents, it was regulated by the Ministry of Health in 1996 that four kinds of diagnostic reagents including HBsAg, anti-HCV, anti-HIV and syphilis should be tested and released by NICPBP on a batch-by-batch basis.

NICPBP is also in charge of the preparation, calibration and distribution of biological standards and reference reagents for manufacturing, research and educational institutions. The unitage of the national biological substances has been calibrated with that of the established international standards or by other methods if they are available. At present, 151 kinds of biological substances have been prepared and distributed.

At present, China is facing some major challenges. Firstly, priority should be stressed on biotechnology to develop our new biologicals; secondly, we should raise our understanding of the importance of biological standardization in the biological field and measures to be taken to put more resources on standardization work. Thirdly, as the regulatory criterion has become more or less international, we should strengthen our relation with WHO collaborating centres for biological and other national control authorities to improve our regulatory environment so as to meet, step by step, the international need. Therefore, strong actions need to be taken by the government to face these challenges to create more possibilities for this development in a speedy and sound way.

Dr. H. Zhou, National Institute for the Control of Pharmaceutical & Biological Products,
N° 2 Tiantan Xili Temple of Heaven, Beijing 100050, China.

Brown F, Griffiths E, Horaud F, Schild GC (eds): A Celebration of 50 Years of Progress in Biological Standardization and Control at WHO.
Dev Biol Stand. Basel, Karger, 1999, vol 100, pp 159-170

International Challenges and Public Policy Issues

N. Morris

Department of Science & Technology Studies, University College London, London U.K.

Key words: Biologicals, regulatory system, standards, quality control, public policy, WHO.

Abstract: The paper presents an overview of current public policy issues relating to biological standardisation and control, drawing on the extensive background material assembled for two recent international reviews, and previously published work. It identifies a number of factors which are destabilising the current system and promoting a climate for change. These include the squeeze on public sector resources, the growth in volume and complexity of biologicals, developing world needs, concerns about harmonisation and new social and ethical issues.
It is argued that this situation presents important opportunities for reviewing the existing boundaries between regulatory scientists, industry, and the public, for international agreement on priorities and for harmonisation and mutual recognition. While considerable progress has already been made on these issues at national, regional and global level, there is a need for fuller international participation and the additional impetus that would come from a higher -profile commitment by governments. Such commitment will also be important for the vital questions of sustaining the scientific base and securing the resource for an effective, truly worldwide programme of standardisation and control. An international approach will also be essential in steering biologicals control through the difficult social and ethical questions of the future. WHO, in collaboration with national authorities, has a key role to play in these developments.

INTRODUCTION

This paper presents an overview of current public policy issues relating to biological standardisation and control. This is a critical time: the current system is under stress from a number of factors. Recent reviews have elucidated what these factors are, and led to reports and proposals about the kinds of adaptation the system may need (see the World Health Organisation (WHO) review group report [1] and the 1997 Review by the UK National Biological Standards Board (NBSB) [2]). The reviews have rightly concentrated on scientific and technical issues, but many critical issues are at the level of public policy, requiring the commitment of governments, the support of the public, and the cooperation of industry and of international institutions.

In the course of this symposium many speakers have highlighted particular policy issues, relevant to the topics they have discussed. This is an opportunity to draw these threads together. In this paper I shall briefly summarise the factors that are forcing change and review the public policy issues arising which call for action at the international level. Discussion of the issues will draw on the extensive background material assembled for the reports mentioned above, and previously published work [3]. The aim is to produce an agenda for consideration by policy-makers who are concerned to maintain and strengthem this important aspect of public health management.

Forces for Change

As background we need to consider the factors that are destabilising the current system for control and standardisation of biologicals on a worldwide scale. The current highly respected international system, in which WHO has played a leading part, is built on the foundation of national institutions. These institutions provide the material resources for the international work and retain sovereignty for the implementation of standards. The system draws its authority from the prestige, integrity and thoroughness of its expert scientific advisers, and relies on the voluntary, and invariably unpaid, cooperation of the national institutions designated as Reference Laboratories and Collaborating Centres for material resource (eg research studies, testing programmes, drafts of documents).

Developments that have been gathering force in the last 10-15 years are now disturbing this:

- squeeze on laboratories acting as Reference Laboratories and Collaborative Centres as governments in the developed world require national institutions to reinvent themselves along lines required by policies of retrenchment in public sector spending and privatisation or partnerships with the private sector. The recent resurgence of the Left in the politics of many western countries may modulate but will not destroy these policies;
- rapid growth in numbers and complexity of biologicals: both the national authorities and the international system cannot keep up with this growth. This is requiring decisions on priorities, delegation (eg to manufacturing industry or third party testing agencies), work-sharing and harmonisation;
- worldwide demand for International Standards and quality control is increasing as the the developing world opens up further as a market and manufacturer of biologicals, raising questions on the financing and training capacity required to meet these needs;
- concerns about harmonisation, ranging from industry's concern about the burden of idiosyncratic differences in national regulatory requirements to public repercussions of conflicting licensing decisions, in a world where news travels increasingly quickly and no nation is an island;
- social and ethical issues, including questions raised by diagnosis and treatment opportunities opened up by some new classes of biologicals (eg genetic diagnosis, xenotransplantation) combined with greater public scepticism over scientific expertise, which require a broadening of the decision-making processes.

I have not in this list referred to the trade issues associated with the international system for biologicals standardisation which have recently been highlighted following the formation of the World Trade Organisation (WTO). These are issues that will require specialist attention, but for the present it appears likely that they can be resolved, temporarily at least, by process changes rather than fundamental reshaping of the current system. Between the role of standards generally in promoting the free flow of goods and yet in specific instances impeding it, and between those agencies which wish everyone to subscribe unreservedly to standards set by WHO and those who consider that a greater degree of national autonomy is essential – the underlying tensions are pervasive and have a bearing on may of the other issues (such as harmonisation and the financing of international standardisation and quality control work).

Issues Arising

The many influences pressing on the system create a scenario for positive change, but also raise causes for concern. Replies to the NBSB and WHO surveys (undertaken for the preparation of the reports referred to earlier) indicate a wish for evolutionary not revolutionary change, and both these and published sources provide substantial evidence of how these changes are already being started.

For purposes of discussion the main opportunities for positive change may be considered under three headings:

- redrawing the boundaries between regulatory scientists, industry, and the public in decision-making and its implementation;
- agreeing priorities on a more representative basis than hitherto;
- harmonisation and mutual recognition – with coordination and rationalisation of the roles of existing national and international institutions.

Of equal importance, but rather more difficult to convert to practical policies, are questions of sustaining the science needed for regulatory control, financing all the activities, and confronting the social, economic and ethical issues that can profoundly affect the working of the system. As an illustration of the concerns and issues arising, the discussion in this paper will highlight the following:

- Scientific basis:
  Keeping the science in the regulation and quality control of biologicals;
  Sustaining expertise and long-term research programmes;
- Resourcing a global programme:
  Resources for the international programme;
  Resources for quality control in developing countries;
- Social and ethical judgments:
  Pace of progress on social and ethical issues;
  The undercurrents of «one standard worldwide»;
  Regulatory borderlines;

Redrawing the Boundaries Between Regulatory Scientists, Industry, and the Public in Decision-Making and its Implementation

Devolution to industry

Developments in science provide the opportunity to move with safety some control responsibilities from National Control Authorities (NCAs) to industry thus helping to control the problems of workload for NCAs without lowering safety standards. There is advantage to industry in that the simplification of external controls speeds up their processes. Scientific advances in two different fields provide the justification for this change. First, the combination of biotechnological methods of manufacture and availability of new and sophisticated methods of physico-chemical analysis is bringing some biologicals closer to «ordinary pharmaceuticals». This gives scope for modifying the regulatory regime for some of these products. Both the EU and the US are moving along this path, (which, in any case, is nothing new, witness vitamins and antibiotics). Secondly, the development and refinement of techniques of risk assessment give a sound basis for relating stringency of regulatory requirement to potential hazard, so that NCA resources are concentrated where risks are judged to be greatest.

Responses to the NBSB questionnaire show widespread support in industry for moving in this direction and also substantial support from the major control authorities. Making this happen, and in an orderly fashion, is an issue for the policy-makers, both inside and outside the biologicals community. They will particularly need, as part of this, to be able to demonstrate that the changes are in the public interest.

Partnership

There is also scope for reshaping the partnership with industry in other ways. The tradition in this field has been one of extreme caution about involvement of industry in its deliberations for fear of 'capture' of the regulatory process by vested interests. This is still a concern. But as the experience of academics entering into research partnerships with industry shows, it is possible to identify a common agenda as the basis for cooperation and to structure relationships so as to manage the diverging or conflicting interests when these arise.

Responses to the NBSB questionnaire showed a notable commitment among industry to working in partnership with regulatory authorities, and the consultations undertaken for the WHO review have drawn attention to the wealth of technical expertise in industry which needs to be drawn on to improve the productivity of discussions.

The complexity and rapid development of the science in the biologicals field, together with the shortage of resource, makes an effective partnership a practical necessity. The issue for public policy however is to manage this partnership without diminishing public confidence in the system. Governments will need to differentiate the kind of partnership to be fostered in this field from that in the politically higher-profile field of innovation, because of the importance of maintaining trust in the independence of regulatory processes.

Partnership with the public

The regulatory system has traditionally been paternalistic with regard to the public. It has also at times used scientific authority as a cloak to shield itself from questioning. Now a better informed and articulate public can more readily pick holes in that cloak – and is not afraid to argue for lay knowledge and lay priorities to be integrated into decision-making. On top of this, in biologicals (and medicine generally), new developments are pushing decision-making more and more into the realms of social and ethical judgments, where lay views have as much or more authority than those of scientists. This trend is discussed further below. For the present, two examples will serve to illustrate the point: what special requirements are needed for (biologicals-based) diagnostic kits that are sold over-the-counter rather than administered under medical supervision? What level of risk is required to justify introducing or retaining a particular resource-demanding screening test for blood?

In general, national governments (and international organisations eg in Europe) are struggling to find effective machinery for public participation. The FDA has an impressive array of opportunities – advertised, *inter alia*, on the WWW – for public participation, but more information is needed on their impact and effectiveness. Although activism on this topic may be largely a Western preoccupation at present, the principle of public participation would seem to be valid worldwide, and important for the success of public health programmes (which require high levels of public confidence) and the acceptability of risk assessments. Different cultures will require different consultative models, but nevertheless this is an area where international cooperation in sharing experience and developing models could be valuable.

PRIORITIES

With so many biological products requiring international quality control, and severe limitations on both human and financial resource, the need for prioritisation is now generally recognised. The challenge is to arrange for more representative input to priority-setting at the international level and to agree the criteria. It becomes a public policy issue and an international challenge because of the large differences between the needs/priorities of different nation states. As one of those (from a developing country) consulted in the preparation of the WHO Review commented:

> *The needs of developed and developing countries should be differentiated. The two environments are worlds apart.*

One of the replies to the NBSB questionnaire put it more starkly:

> *«Governments and people in the developing countries will not be able to afford to implement some of the dramatic new advances that are likely to appear in biologicals.»*

The recommendations of the WHO Review Team are highly relevant to this question, covering as they do wider consultation and opportunities for input, and putting the focus on products of wide application such as vaccines and blood and its products. The public policy issues for debate in individual nation states will be the extent to which national priorities may be subject to compromise or reconsideration or rearrangement in the light of a global policy.

Harmonisation and Rationalisation

This is a big issue, where a number of initiatives are already in progress. The biologicals regulatory industry has grown in volume and complexity along with the products it regulates, and the expectations of a concerned and interested public. WHO has played an important coordinating role, but it is not itself a regulatory body, and its guidance does not have comprehensive coverage nor necessarily universal acceptance. (Hence the concern felt by some nation states when it appeared that WTO procedures would effectively confer on WHO guidance a greater authority than it had hitherto claimed.) The situation that results from this light and somewhat distanced form of coordination is that there are still too many instances where different countries have differing requirements for, eg evidence of safety of a new product, or where different bodies are unnecessarily duplicating each others' work. This is not just a matter of it being tough for industry. It relates back to the problems of workload for NCAs. For both those with sophisticated facilities and those struggling to set themselves up in the developing world, the volume of work to be done can only be controlled by a sensible system of mutual recognition and acceptance of approvals given elsewhere. The public interest is also involved insofar as it is important to avoid unnecessary delays in bringing safe and effective biological products to market. (Note that the public voice – in the shape of AIDS activists – has been influential in bringing about fast track approval systems for therapeutically important drugs.)

Harmonisation, mutual recognition, and agreement between institutions on roles and division of work will need a multi-faceted approach. Much can be done within the biologicals community [4], but substantive progress is likely to require the support or even the involvement of governments. The International Conference on Harmonisation (ICH) has achieved a great deal; so, in a different way has the European Union (EU). There is much to be learnt from their processes. The challenges of a fully global system of harmonisation have however yet to be met. There is undoubtedly a role for WHO but it will be difficult to play even a coordinating and catalytic role on a shoestring. Likewise WHO might be the body to bring together the different national and international standard-making and investigative organisations, to preside over discussions of defining individual, non-overlapping roles – but this activity will also require resourcing.

Mutual recognition has achieved some notable advances on a bilateral basis. Not only has the EU established a system of full mutual recognition among its member states, but also agreements are being put in place with the US and Canada. The EU experience has however demonstrated the sensitivity of perceived differences in standards between authorities in different states, and the need for intensive discussion and mutual assistance to arrive at sensible and trusted ways of working. Difficulties of this sort will arise at the global level too. But basically there is an opportunity here to seize. The sort of piecemeal, incremental progress that has been shown feasible should continue, but the importance of the problem merits a full scale global policy initiative, with full governmental involvement (perhaps as for biodiversity, or anti-pollution). That is perhaps something that the institutions represented here might consider working for.

Scientific Basis

A science-based system

Science-based decision-making continues to be central to biological standardisation and control, but there are concerns arising from the prevailing financial stringency (leading to lay-offs of scientific staff in some agencies) and the trend to more public participation in decision-making. Moreover the changes that are currently being proposed to the machinery for WHO, particularly the ECBS, could be read as opening it up more to political or other considerations. But such fears are probably ungrounded while the nature of the matters under discussion continues as now to make a powerful case for science. The rapid developments in new and improved products are throwing up questions of such scientific and technical complexity and novelty that a high loading of scientific expertise is essential for their resolution. And the high esteem in which the existing expert-based system is held, and its excellent track record in reaching sound and internationally acceptable conclusions, should ensure strong institutional support for conservation of its basic principles of science-based decisions. The challenge is rather one of modernising ways of working to ensure continuing effectiveness in a changing environment – without, as they say, throwing out the baby with the bathwater.

Sustaining expertise and long-term research programmes

A more serious threat to the scientific basis of standardisation and control is the uncertainty surrounding the maintenance of specialist expertise in the longer term. This is likely to become an international challenge because the evidence is that the national institutions which have nurtured the specialism and maintained long-term research programmes hitherto may be under difficulty to do this in the future. This is, once again, a consequence of growth in volume and complexity of work, restructuring of public sector institutions or their funding, combined in this case with a relative lack of interest or capability in academia to provide advanced training. Some institutions have already reduced or abandoned their long-term research: for example, the Danish Statens Serum-Institut has pulled out of Standards-making altogether and handed over to the National Institute for Biological Standardisation and Control (NIBSC) in the UK. There must also be a question mark over how far fully self-funding institutions (like the Therapeutic Goods Administration in Australia) will be able to sustain programmes of research in the future. Some replies to the NBSB questionnaire suggested that new institutions were needed to secure the scientific basis for the future: these would need to be cross-sectoral (ie both government and industry-supported) and also international, to draw together sufficient resource and expertise. For example one reply from industry referred to: *«the possibility of new alliances and structures being evolved to carry out such activities»*, while another from the public health sector suggested: *«Not for profit research based organisation jointly funded from industrial and public sources are an option.»* Given the daunting array of new scientific and technical questions raised by new developments in biologicals (including some of immediate and worldwide health importance, such as combination vaccines) ensuring the maintenance of scientific capability is properly a concern of policy-makers at the highest national and international level and may call for imaginative approaches.

Resourcing a Global Programme

Resources for the international programme

Whereas the sustainability of research and training in the long term is one particular issue that relates to resourcing, there are also resource considerations concerning the immediate programme of work on International Standards and other quality control measures. WHO has the status to lead an international programme, but does not control the resources that this requires. This is not necessarily a bad thing: cooperative agreements which allow for local control of finances can be a less bureaucratic and more effective system than centralised funding. But the evolution of the present international system has led to imbalance. The situation regarding International Standards is a glaring example. At present about 95% of new Standards and the vast majority of existing ones emanate from a single national establishment. It is a public-funded institution – a status it has maintained after a recent challenge over privatisation because of the public health importance and need for independence in its work. This leads to the odd situation that the implementation of the entire world system of International Standards for biologicals comes courtesy of an unsolicited gift from a single national government. Organisations or businesses that depend on International Standards to carry out their work effectively and safely might well feel very uneasy with this situation. Not only might there be fears that the supply could be abruptly cut off, but also those who are sceptical of altruism might worry about whether such a relationship might skew priorities.

In seizing the opportunities referred to above for harmonisation, mutual recognition and work sharing, the whole question of resourcing also needs to be addressed. There are likely to be advantages in concentration of expertise, and different institutions having different specialisms. But a strong and flexible international system for the future will require partnerships and inter-dependency and an agreed financial strategy for the system as a whole. This may require some additional resource for coordination and change management, but need not depart from the principles of devolved and item-specific funding.

Resources for quality control in developing countries

The current basis for the implementation of quality standards is that this is the responsibility of national authorities. But, in the words of one of those from developing countries replying the WHO Review Team's recent consultation:

> *We should not forget that in many developing countries the national control authorities do not exist or exist nominally.*

Thus an essential part of the international programme for standardisation and quality control of biologicals is the strengthening or establishment of national authorities. The respondent quoted above goes on to say:

> *WHO should first help these countries by training the qualified personnel of such countries to be able to control properly the biological products.*

Another respondent from a developing country refers to:

> *The need to develop a more intense work in the harmonisation among countries or groups of countries, mainly those less developed, in order to poise the stringency level of the regulatory authorities.*

Such assistance is also given priority by respondents from agencies in developed countries, and by industry.

The resource problems of achieving this do however require further consideration. To some extent WHO has been able to rely on funds donated or budgeted for particular disease programmes to fund work on establishing or improving national control systems, but this can lead to fragmentation and the level of funding may not be adequate. There are thus policy issues, for developed countries in particular, as to how far they are prepared to contribute resource in this cause. The issue is not just one of financial resource – or not directly. A prime need is for training, i.e. human resource. The knowledge and expertise required resides mainly in the staff of agencies in the developed countries. The question is whether the governments of these countries will accept the need to maintain sufficient capacity in their agencies to contribute to this training function. This applies whether they support the agencies directly through public funds or deal with them at arms length through a self-funding system. Such training can take many forms: consultancy, secondments (in both directions), and networking between laboratories.

A related question (alluded to in the second of the quotations above) is whether the emphasis should shift to regional rather than national authorities. This is an idea that has been around for many years, but there has not been substantial progress in this direction. The most notable success has been in Europe, deriving from the economic union of European Union member states. The cynic might note that the «regionalisation» of biologicals' control in Europe (like European harmonisation measures in other fields) has been carried through in the name of facilitating the free movement of goods and services between member states, rather than in the name of public health or efficient use of scarce skills. Initiatives in other areas (South America for example) are making progress, but there are still difficulties to resolve. The question of whether policy should focus on national, or regional authorities, or some system of networking which tried to have both, is one worth further airing.

Social and Ethical Judgments

Pace of progress on social and ethical issues

I have already referred (in discussing «Partnership with the Public») to some of the social and scientific developments that are bringing decisions on biologicals more into the arena of social and ethical debate. While developing that partnership is likely to be one of the policy responses to this situation, there are some areas where separate advisory machinery may be needed to explore ethical and social issues. A particularly sensitive area is therapy and diagnosis for genetic conditions. This raises questions ranging from legal rights to employment prospects to eligibility for insurance to religious belief as well as medical and public health issues. The process of bringing some new biologicals to market will in future depend on the pace of progress on these non-scientific and medical issues. This adds further complexity to the task of prioritisation of work on standards and guidelines, and will require further work on establishing appropriate relationships with these complementary advisory and decision-making bodies.

The meaning of «one standard worldwide»

The whole international system of Standards and quality control is based on the principle of a single standard worldwide. In the case of International Reference Standards, this principle is unassailable. On the other hand, where quality control is concerned, the statement of principle may be subject to a host of silent reservations. The area abounds with political, economic and ethical sensitivities. Because the notion that poorer countries should have lower standards of quality for their biological products is offensive, this tends to suppress discussion of whether the current western-dominated standards are appropriately pitched. It is hard to disentangle ethical and humanitarian principles from vested interests in the established pharmaceutical industry's strong commitment to stringent, universally applied quality standards. Meanwhile we know that there is a desire to set up local manufacture of biologicals in parts of the developing world. Concerns about whether they will all be able to meet the quality standards are quite frequently expressed, but the implications of this are not fully explored. There are 4 options:

- no local manufacture – country goes without or possibly has aid to import;
- aid to bring local products up to standard;
- the «blind eye» – everyone pretends the differences are not there;
- move the goalposts – waive or adapt quality control standards.

Although the last option comes in the realm of the unacceptable, it should not therefore be undiscussed, if only because it serves to focus discussion of the other options. It would also be an opportunity to acknowledge more clearly the scope for flexibility in such standards. NCAs, including those in developed countries, can argue the equivalence of their own requirements to those agreed by WHO. Moreover, a growing number of NCAs are now relying on risk assessment systems to determine the level of quality control of particular products in the light of judgments about the risk to their population. The single-standard principle is a good principle, but still needs from time to time to be evaluated and interpreted against the real world, and the appropriateness of quality control measures to specific environments.

Regulatory borderlines

We are accustomed to biologicals evolving over a period to «ordinary» pharmaceuticals, and pharmaceuticals themselves being liberated from the status of «ethicals» to «over-the-counter» preparations. There may also be enlargement of the boundaries to incorporate new kinds of product – such as organs for transplant. The process is ill-defined, and there is some arbitrariness about what is included as a biological (or a medicine for that matter). For example human organs for transplant are most often not subject to biologicals regulation, but animal organs for transplant, and possible the transgenic donor animals themselves, are being drawn into the net.

The question of what is, and is not, a medicine seems poised to become more difficult in the future. The techniques of biotechnology are enabling the modification of food and cosmetics to perform quasi-medical functions [5]. While a vaccine delivered in a banana [6] will clearly be treated as a biological medicine, other more borderline products can be expected to appear in the future and be marketed as foodstuffs, as has already happened for pharmaceuticals. The anomalies likely to arise may also call into question other divides that are currently accepted, such as the freedom from regulation of medicines deriving from traditional systems outside the Western medical culture, but now enjoying a renewed respect and popularity. These matters will constitute some of the more difficult policy issues for the future.

Conclusion

This has been a brief overview, highlighting just some of the issues beyond the scientific and technical, which are likely to exercise the biologicals community – along with governments and the public that both endeavour to serve – in the immediate future. They are international challenges not only because biological medicine use and manufacture has been trans-national throughout the history of organised standardisation and control, but also because the increasing demands on the system make international cooperation and work-sharing essential in the public interest. As well as the new demands arising from scientific and technical advances, there are ambitious programmes of public health sponsored by WHO and other bodies. The pressure of these demands, along with the key issue of harmonisation, demonstrates the need for a more international approach to repair the discrepancies or duplications that have arisen from a fragmented national or regional set of systems. The issues concerned with readjusting relationships with industry and with a better-informed public will benefit from an international approach to determining policy. This will serve both to share experience and to limit the confusion and disaffection that results from the ripple effect of decisions of one country on communities elsewhere. Similar arguments apply to the social and ethical issues that are becoming more and more woven in to decisions and priorities in biologicals' standardisation and control. Finally, there can be no shirking the issue of resource. While the bulk of spending will continue to be by individual nation states on their own regulatory affairs, it will be essential also to have international agreement on funding: for coordination and the important tasks carried out through the ECBS; for assistance and training for quality control functions in developing countries: and for assuring the long-term maintenance of expertise and programmes of research in the public interest. The biologicals community, as represented at this meeting, is in a position of influence which can help create the conditions in which these international challenges can successfully be met.

Acknowledgements

I am grateful to the UK NBSC and to the WHO for permission to use material gathered in my capacity as consultant in the course of their reviews and to Dr JC Petricciani and Dr GC Schild for advice generously given.

References

1 Petricciani JC, Chamone D, Vicari G: Review of remit and activities of WHO in the biologicals field and the Biologicals Unit. Report presented to WHO in October 1998.

2 Biological Standardization and Control: a scientific review commissioned by the UK National Biological Standards Board (NBSB). Geneva, World Health Organisation, 1997.

3 Morris N: Biological medicines in the age of biotech: public policy issues. Science and Public Policy 1997;24(1):53-62.

4 Petricciani JC: International expectations for harmonization of biological standards: Food and Drug Law Journal 1994;49:397-400.

5 Enriques J: Genomics and the world's economy. Science 1998;281:925-926.

6 Tacket CO, Amtzen CJ: Nature Med 1997;4:607.

Norma Morris, Department of Science & Technology Studies, University College London, Gower Street, London WC1E 6BT, U.K.

List of Participants

ANDERLE K., Baxter/Immuno, Industriestrasse 72, A-1220 Vienna.

ANDRE F.E., SmithKline Beecham Biologicals, 89, rue de l'Institut, B-1330 Rixensart.

ARITA I., ACIH-Agency for Cooperation in International Health, 4-11-1 Higashi-machi, Kumamoto City, Kumamoto 862-0901, Japan.

ARMSTRONG D., Natal Bioproducts Institute, 10 Eden Road, Pinetown 3610, Private Bag X9043, Pine Town, 3600, South Africa.

ASHER D.M., Center for Biologics Evaluation & Research, FDA, Laboratory of Development, Division of Viral Products, 1401 Rockville Pike-HFM-255, Rockville, MD 20852-1448, USA.

BANGHAM D.R., 4, Crown Close, Mill Hill, London NW7 4HN, U.K.

BARON H., Centeon S.A., 15, rue de la Vanne, F-92945 Mont-Rouge.

BOSTVIRONNOIS C., Pasteur Mérieux Connaught, 58, avenue Leclerc, F-69007 Lyon.

BOUDIER A., Biovector Therapeutics, Chemin du Chêne Vert, B.P. 169, F-31676 Labège Cedex.

BROWN F., Plum Island Animal Disease Center, P. O. Box 848, Greenport, NY 11944, USA.

CHARIATTE N., S.S.I.V., Rehhagstrasse 79, P.O.Box, 2707, CH-3001 Berne.

CHUMAKOV K., Center for Biologics Evaluation & Research, FDA Division of Viral Products, 1401 Rockville Pike, HFM-255, Rockville, MD 0852-1448, USA.

CICHUTEK K., Paul-Ehrlich Institute, Department of Medical Biotechnology, Paul-Ehrlich Strasse 51-53, D-63255 Langen.

DOBBELAER R., Scientific Institute of Public Health, Wytsmanstrasse 14, B-1050 Brussels.

ECKHARDT T., Wyeth-Lederle Vaccines, Middletown Road, Pearl River, NY 12553, USA.

EGAN W., Office of Vaccines Research and Review, CBER/FDA, 1401 Rockville Pike, Rockville, MD 20852-1448, USA.

ESHKOL A., Ares Serono International, 15, chemin des Mines, CH-1202 Geneva.

EVERS T., European Plasma Fractionation Association (EPFA), Plesmanlaan 125, NL-1006 AD Amsterdam.

FITZGERALD E., Fitzgerald Consulting, 3, Owens Court, Rockville, MD 20850, USA.

FUMERO S., Ares Serono International, 15, ch. Des Mines, CH-1202 Geneva.

GAUDRY D., c/o M.-C. Garnier, Mérial, 29, av. Tony Garnier, B.P. 7123, F-69348 Lyon Cedex 07.

GRACHEV V., Chumakov Institute of Poliomyelitis, PO Institute of Poliomyelitis, Moscow Region 142782, Russia.

GRIFFITHS E., World Health Organization, Biologicals Unit, Av. Appia, CH-1211 Geneva 27.

GUST I., C S L Ltd, 45 Poplar Road, Parkville, Victoria 3052, Australia.

HAYAKAWA T., National Institute of Health Sciences, Biological Chemistry and Biologicals, 1-18-1 Kamiyoga, Setagaya-Ku, Tokyo 158-8501, Japan

HILLEMAN M.R., Merck Institute for Therapeutic Research, Merck Research Laboratories (WP53C-35C), Sumneytown Pike, West Point, PA 19486, USA.

HORAUD F., Institut Pasteur – Texcell, 28, rue du Dr Roux, F-75724 Paris Cedex 15.

HOWALD M., Swiss Federal Veterinary Office, Schwarzburgstrasse 161, CH-3003 Berne.

HULSE E.C., 12 Beacon Road, Ditchling, Hassocks, W. Sussex BN6 8UL, U.K.

HULSE C., British Ambassy, Thunstrasse 50, P.O. Box, CH-3000 Berne.

JEFFCOATE S., P.O. Box 188, UK-Dorking, Surrey RH4 2FE, U.K.

KIM H.-S., Boran Pharmaceuticals, 437-3 Ahyun-Dong, Mapo.Gu, Seoul 121-010, Korea

LELIE N., CLB, Sanguin Blood Supply Foundation, Viral Diagnostic Laboratory, Plesmanlaan 125, NL-1066 CX Amsterdam.

LOEWER J., Paul-Ehrlich Institut, Human Virology, Paul-Ehrlich Strasse 51-57, D-63225 Langen.

LUBINIECKI A., SmithKline Beecham Pharmaceuticals, 709 Swedeland Road, King of Prussia, PA 19406, USA.

MARTIN J.-F., Biocine, 36, quai Fulchiron, F-69005 Lyon.

MATSUKI S., Kirin Brewery Co., Ltd, CJPMA, 100-1 Hagiwara-Machi, Takasaki, Gunma 370-0013, Japan.

MEIGNIER B., Pasteur Mérieux Connaught, 1541, avenue Marcel Mérieux, F-69280 Marcy l'Etoile.

MERIEUX C., Fondation Marcel Mérieux, 17, rue Bourgelat, F-69002 Lyon.

MILLS K., Maynooth University, Infection & Immunity Group, St Patrick's College, Maynooth Co. Kildare, Ireland.

MIRE-SLUIS A., NIBSC, Blanche Lane, South Mimms Potters Bar, Herts EN6 3QG, U.K.

MONTAGNON B., 5, rue des Ecoles –Le Bourg, F-69210 Lentilly.

MORIN-CARPENTIER O., IFPMA, 30 rue de St-Jean, CH-1211 Geneva 18.

MORRIS N., University College London, Science & Technology Studies, Gower Street, London WCIE 6BT, U.K.

MURANO G., U.S. Pharmacopeia, 12601 Twinbrook Parkway, Rockville, MD 20852, USA.

OESCH B., Prionics AG, University of Zürich, Winterthurstrasse 190, CH-8057 Zürich.

OUTSCHOORN A., US Pharmacopeia, 12601 Twinbrook Parkway, Rockville, MD 20852, USA.

OTTIGER H.P., Institut für Viruskrankenheiten und Immunoprophylaxe, Postfach, Sensemattstrasse 293, CH-3147 Mittelhäusern.

PADILLA A., W.H.O, Biologicals Unit, 20 avenue Appia, CH-1211 Geneva.

PALYA V., Sanofi Phylaxia Veterinary Biologicals Ltd, Szigetvari U 10, H-1222 Budapest.

PEETERMANS J., SmithKline Beecham Biologicals, 89, rue de l'Institut, B-1330 Rixensart.

PETRICCIANI J., P.O. Box 1925, Palm Springs, CA 92263, USA.

POLACEK M., Veteran Affairs Dept, 12286 Old Country Road, Wellington, FL 33414, USA.

REIGEL F., Federal Office of Public Health, Hess Strasse 27e, CH-3003 Berne.

ROBERTSON J.S., NIBSC, South Mimms, Blanche Lane, Potters Bar, Herts EN6 3QG, U.K.

RONCHI E., OECD, 7, rue André Pascal, F-75016 Paris.

ROUMIANTZEFF M., 1 rue Dangon, F-69004 Lyon.

SCHILD G., NIBSC, South Mimms, Blanche Lane, Potters Bar, Herts EN6 3QG, U.K.

SESARDIC T., NIBSC, South Mimms, Blanche Lane, Potters Bar, Herts EN6 3QG, U.K.

SHAW A., Merck Research Laboratories, Virus & Cell Biology, 770 Sumneytown Pike, Bldg 16 Room 100, West Point, PA 19486-0004, USA.

SHIN S.H., Boran Pharmaceuticals, 3FL, Koryo Akademitel, 437-3 Ayunn-Dong, Mapo-Gu, Seoul 121-010, Korea.

SIBER G., Wyeth-Lederle Vaccines & Pediatrics, 401 North Middletown Road, Pearl River, NY 10965, USA.

SPIESER J.-M., Conseil de l'Europe, Service Européen de la Qualité du Médicament, B.P. 907, F-67029 Strasbourg Cedex 1.

STRUCK M.M., Swiss Serum and Vaccine Institute, Rehhagstrasse 79, CH-3018 Berne.

THOMAS D.,The Old Barn, North Green, Kirtlington, Oxford OX5 3JZ, U.K.

VAN AKEN W.G., CBL, Medical Director, Plesmanlaan 125, NL-1066 CX Amsterdam.

VAN DE DONK H., R.I.V.M., P.O. Box 1, NL-3720 BA Bilthoven.

VICARI G., Istituto Superiore di Sanita, Viale Regina Elena 299, I-00161 Roma.

VYAS Gi.N., University of California School of Medicine, UCSF Box 0134, Department of Laboratory Medicine, San Francisco, CA 94143, USA.

WOOD D.J., NIBSC, South Mimms, Blanche Lane, Potters Bar, Herts EN6 3QG, U.K.

ZHOU H.-J., Ministry of Health, National Institute for the Control of Pharmaceutical and Biological Products, N° 2 Tiantan Xili Temple of Heaven, PRP-Beijing 1000500, China.

ZOON K, FDA/CBER, 1401 Rockville Pike, Suite 200N, Rockville, MD 20852, USA.

ZORZI-MORRE P., Sanofi Recherche, Développement en Biotechnologie, Centre de Labège, Bureau N-218, B.P. 137, F-31676 Labège Cedex.

Index of Authors

Symposia Series in Immunobiological Standardization

Vol. 1: Talloires 1965	Rabies
Vol. 2: Munich 1965	Neurovirulence of viral vaccines
Vol. 3: Stockholm 1965	Biotechnical developments in bacterial vaccine production
Vol. 4: Royaumont 1965	Immunological methods of biological standardization
Vol. 5: London 1966	Laboratory animals
Vol. 6: Utrecht 1966	Adjuvants of immunity
Vol. 7: Marburg/L. 1967	Assay of combined antigens
Vol. 8: Lyon 1967	Foot-and-Mouth Disease: Variants and immunity
Vol. 9: Paris 1967	Pseudotuberculosis
Vol. 10: London 1967	Biological assay methods of vaccine and sera
22nd Symp., London 1968	Standardization of immunofluorescence. Ed. Blackwell
Vol. 11: London 1968	Rubella vaccines
Vol. 12: Tunis 1968	Brucellosis. Standardization and control of vaccines and reagents
Vol. 13: Utrecht 1969	Pertussis vaccine
Vol. 14: London 1969	Standardization of interferon and interferon inducers
Vol. 15: Berne 1969	Enterobacterial vaccines
Vol. 16: Versailles 1970	Anti-lymphocytic serum
29th Symp., Brighton 1969	Hazards of handling simians. Ed. Lab. Animals Ltd. (4)
Vol. 17: Frankfurt 1970	BCG vaccine
Vol. 18: Copenhagen 1972	HL-A reagents
Vol. 19: Utrecht 1972	Smallpox vaccines
Vol. 20: London 1972	Influenza
Vol. 21: Lyon 1972	Rabies II
Vol. 22: Monaco 1973	The use of vaccines

Developments in Biological Standardization

Vol. 23: Madrid 1973	Sterilization and sterility testing of biological substances
Vol. 24: San Francisco 1973	Preservatives in biological products
Vol. 25: Lyon 1973	Requirements for poultry virus vaccines
Vol. 26: Budapest 1973	Selected veterinary vaccines
Vol. 27: Budapest 1973	Purity of human plasma proteins
Vol. 28: London 1974	Immunity to infections of the respiratory system in man and animals
46th Symp., London 1974	Breeding of simians and their uses in developmental biology. Ed. Lab. Animals Ltd. (6)
Vol. 29: Geneva 1974 (WHO/IABS)	Standardization and control of allergens administered to man
Vol. 30: Milan 1974	Viral hepatitis
Vol. 31: Rabat 1975	Brucellosis (II)
Vol. 32: Paris 1975 (OIE/IABS)	Colstridial sera and vaccines for veterinary use
Vol. 33: Douglas 1975	Vaccination by the non-parenteral route
Vol. 34: Budapest 1976	Pyrogenicity, innocuity and toxicity test systems for biological products
Vol. 35: Lyon 1976	Foot-and-mouth disease vaccines

Vol. 36: Lyon Washington D.C. 1976	Freeze-drying of biological products
Vol. 37: Geneva 1976	Standardization of cell substrates for the production of virus vaccines
Vol. 38: London 1977	Biological preparations in the treatment of cancer
Vol. 39: Geneva 1977	Influenza immunization (II)
Vol. 40: Marburg 1977	Standardization of rabies vaccines (III) (WHO/IABS)
Vol. 41: Guadeloupe 1978	Vaccinations in the developing countries
Vol. 42: Paris 1978	Second General Meeting of European Society of Animal Cell Technology
Vol. 43: Brussels 1978	Immunization: Benefit versus risk factors
Vol. 44: Villars 1979	Test methods for the quality control of plasma proteins
Vol. 45: San Antonio 1979	The Standardization of animals to improve biomedical research, production and control
Vol. 46: Oxford 1979	Third General Meeting of European Society of Animal Cell Technology
Vol. 47: Bilthoven 1980	Reassessment of inactivated poliomyelitis vaccine
Vol. 48: Geneva 1980	Standardization of albumin, plasma substitutes and plasmapheresis
Vol. 49: Leetown 1981	Fish biologics: Serodiagnostics and vaccines
Vol. 50: Heidelberg 1981	Joint ESACT-IABS Meeting on the use of heteroploid and other cell substrates for the production of biologicals
Vol. 51: Lyon 1981	Immunization of adult birds with inactivated oil adjuvant vaccines
Vol. 52: Lyon 1981	Herpes virus of man and animal: Standardization of immunological procedures
Vol. 53: Dublin 1982	Enteric infections in man and animals: Standardization of immunological procedures
Vol. 54: Athens 1982	Viral hepatitis: Standardization in immunoprophylaxis of infections by hepatitis viruses
Vol. 55: Copenhagen 1982	Fifth General Meeting of European Society of Animal Cell Technology
Vol. 56: Algeria 1983	IIIrd International Symposium on Brucellosis
Vol. 57: Paris 1983	Monoclonal antibodies: Standardization of their characterization and use
Vol. 58: Budapest 1983	BCG vaccines and tuberculins
Vol. 59: Geneva 1983	Standardization and control of biologicals produced by recombinant DNA technology
Vol. 60: Gardone Riviera 1983	Production and exploitation of existing and new animal cell substrates
Vol. 61: Geneva 1984	Proceedings of the Fourth International Symposium on pertussis
Vol. 62: Stockholm 1985	Diagnostics and vaccines for parasitic diseases
Vol. 63: San Francisco 1985	Use and standardization of chemically defined antigens
Vol. 64: London 1985	Reduction of animal usage in the development and control of biological products
Vol. 65: Amsterdam 1985	Use and standardization of combined vaccines
Vol. 66: Baden 1985	Advances in animal cell technology: cell engineering, evaluation and exploitation
Vol. 67: Melbourne 1986	Standardization in blood fractionation including coagulation factors
Vol. 68: Geneva 1986	Cells – Products – Safety
Vol. 69: London 1987	Cytokines: Laboratory and clinical evaluation
Vol. 70: Arlington, 1988	Continuous cell lines as substrates for biologicals
Vol. 71: Utrecht, 1989	Monoclonal antibodies for therapy, prevention and *in vivo* diagnosis of human disease
Vol. 72: Annecy, 1989	Progress in animal retroviruses
Vol. 73: Shizuoka, 1990	Pertussis: Evaluation and research on acellular pertussis vaccines
Vol. 74: Bethesda, MD, 1990	Biological product freeze-drying and formulation
Vol. 75: London, 1990	Virological aspects of the safety of biological products
Vol. 76: Bethesda, MD, 1991	Continuous cell lines – Current issues
Vol. 77: Annecy, 1991	Standardization of the immunopharmacology of natural and synthetic immunomodulators
Vol. 78: Bethesda, MD, 1991	Poliovirus attenuation: Molecular mechanisms and practical aspects
Vol. 79: Ploufragan, 1992	The first steps towards an international harmonization of veterinary biologicals: 1993 and free circulation of vaccines within the E.E.C.
Vol. 80: Heidelberg, 1992	Transmissible spongiform encephalopathies-Impact on animal and human health.
Vol. 81: Cannes, 1992	Virological safety aspects of plasma derivatives.
Vol. 82: Albany, NY, 1993	Recombinant vectors in vaccine development.
Vol. 83: Annecy, 1993	Genetic stability and recombinant product consistency.

Vol. 84: Langen, 1993	Non-target effects of live vaccines.
Vol. 85: Santa Fee, NM, 1994	Genetics of streptococci, enterococci and lactococci.
Vol. 86: Langen, 1994	Replacement, reduction and refinement of animal experiments in the development and control of biological products.
Vol. 87: Geneva, 1995	New approaches to stabilisation of vaccines potency.
Vol. 88: Bethesda, MD, 1995	Viral safety and evaluation of viral clearance from biopharmaceutical products.
Vol. 89: Rome, 1995	Pertussis vaccine trials.
Vol. 90: Oslo, 1996	Fish vaccinology.
Vol. 91: San Francisco CA, 1996	Development of specifications for biotechnology pharmaceutical products.
Vol. 92: Bergen, 1996	Modulation of the immune response to vaccine antigens.
Vol. 93: Annecy, 1996	Safety of biological products prepared from mammalian cell culture.
Vol. 94: Bethesda, MD, 1997	Simian virus 40 (SV40): a possible human polyomavirus.
Vol. 95: Paris, 1997	Preclinal and clinical development of new vaccines.
Vol. 96: Washington DC, 1995	Characterization of biotechnology pharmaceutical products.
Vol. 97: London, 1997	Biological characterization and assay of cytokines and growth factors.
Vol. 98: London, 1997	Inactivated influenza vaccines prepared in cell culture.
Vol. 99: Strasbourg, 1998	Animal sera, animal sera derivatives and substitutes used in the manufacture of pharmaceuticals: Viral safety and regulatory aspects.

International Association of Biological Standardization

What is the I.A.B.S. ?

The International Association of Biological Standardization (IABS) was founded in Lyons in 1955 by a group of independent experts who identified an urgent need for an improvement in the quality and comparability of the data being exchanged between scientists working in the research, development, production, standardization and regulation of biological products. To this end, IABS aims to bring together state controllers, manufacturers and research workers from academic institutions and public health organisations.

The IABS is a branch of the International Union of Microbiological Societies and has about 500 members in 50 countries. Every four years, its members elect the Council by postal ballot and its representatives, in their turn, elect the officers.

The duty of the Council is to organise meetings on subjects of contemporary interest in biological products being used in human and veterinary medicine and to promote the editorial activity of the IABS by publishing: i) the proceedings of the international meetings in the symposium series *Developments in Biological Standardization* published by S. Karger, ii) *Biologicals,* a major international journal published quarterly by Academic Press (London), and iii) the *Newsletter* which reports on current IABS activities.

The Council also supports the objectives of the major international public health programmes and organisations, and particularly the World Health Organization.

In 1994, the Council decided to establish an IABS Task Force on Vaccines to provide a focus for our activities in the field of vaccination and immunisation. The role of the Task Force is to foster communication between vaccine manufacturers, scientific research workers and regulatory authorities concerning the quality control, standardization and efficacy of vaccines and vaccine combinations and the exchange of information on the development of new improved vaccines. A major objective of the Task Force will be the organisation of international meetings and the promotion of high quality publications on vaccines in our journal *Biologicals* and in the *Developments in Biological Standardization*.

The Task Force has a membership of 30 leading international experts on vaccination and immunisation working together to identify the major topics in the field of vaccines that need international debate and cooperation.

IMPRIME
RIE MEDE
CINE m+h
HYGIENE
octobre-1999